Introduction to Oil Tanker and Gas Carrier Operations

Introduction to Oil Tanker and Gas Carrier Operations introduces the mandatory minimum requirements for training and qualifications for masters, officers and ratings serving on-board liquefied gas tankers. It covers basic safety and pollution-prevention precautions and procedures, layouts of several types of liquefied gas tankers, types of cargo, their hazards and their handling equipment, as well as general operational sequence and liquefied gas tanker terminology.

The book is intended for officers and key ratings who have not previously served onboard crude oil or liquefied gas tankers as part of the regular ship's company. It covers the Level 1 training requirements prescribed by Regulation V/1, paragraph 1.2 of the International Convention on Standards of Training, Certification and Watchkeeping for Seafarers, STCW-95.

Alexander Arnfinn Olsen is a training solutions architect for a global maritime engineering company and Senior Consultant at RINA Consulting Defence, UK. He is STCW II 1995 qualified and has also worked as a marine training designer, marine auditor and fisheries observer. His other books with Routledge include *Introduction to Container Ship Operations and Onboard Safety*, *Maritime Cargo Operations* and *Merchant Ship Types*.

Introduction to Oil Tanker and Gas Carrier Operations

Alexander Arnfinn Olsen

Routledge
Taylor & Francis Group
LONDON AND NEW YORK

Front cover image: bob63/Shutterstock

First published 2025
by Routledge
4 Park Square, Milton Park, Abingdon, Oxon OX14 4RN

and by Routledge
605 Third Avenue, New York, NY 10158

Routledge is an imprint of the Taylor & Francis Group, an informa business

British Library Cataloguing-in-Publication Data
A catalogue record for this book is available from the British Library

ISBN: 978-1-032-82400-0 (hbk)
ISBN: 978-1-032-82541-0 (pbk)
ISBN: 978-1-003-50504-4 (ebk)

DOI: 10.1201/9781003505044

Typeset in Sabon
by Newgen Publishing UK

Contents

Preface

In the late 1920s, the transportation of liquefied gases in bulk began. In the very beginning it was the transportation of propane (C_3H_8) and butane (C_4H_{10}) in fully pressurised tanks. Around 1959, semi-pressurised ships entered the market and liquefied gas was now transported under lower pressure, which was made possible by lowering the temperature. By 1963, fully refrigerated ships for LPG, LNG and certain chemical gases wore in service, carrying cargo at atmospheric pressure. Liquefied gas is divided into distinct groups based on boiling point, chemical bindings, toxicity and flammability. The distinct groups of gases have led to several types of gas carriers and cargo containment system for gas carriers. The sea transport of liquefied gases in bulk is internationally regulated – regarding safety through standards established by the International Maritime Organisation (IMO), and these standards are set out in the IMO's Gas Carrier Codes, which cover design, construction and other safety measures for ships carrying liquefied gases in bulk.

This book is intended for officers and key ratings that have not previously served onboard liquefied gas tankers as part of the regular ship's company. It covers mandatory minimum training requirements prescribed by Regulation V/1, paragraph 1.2 of the International Convention on Standards of Training, Certification and Watchkeeping for Seafarers, STCW-95, and includes basic safety and pollution-prevention precautions and procedures, layouts of several types of liquefied gas tankers, types of cargo, their hazards and their handling equipment, general operational sequence and liquefied gas tanker terminology.

The purpose of this book is to introduce the reader to the mandatory minimum requirements for training and qualifications for masters, officers and ratings serving onboard liquefied gas tankers.

Alexander Arnfinn Olsen
May 2024

Chapter 1

History of the oil and chemical bulk shipping industry

GENERAL

The technology of oil transportation has evolved in parallel with the oil industry itself. Although the use of oil as fuel reaches far back to prehistory, the first modern commercial exploitation of oils distilled from material extracted from the ground is credited to the Scottish chemist, James Young, who first manufactured paraffin (kerosene) in 1850 from coal and oil shales. In the early part of the 1850s, oil began to be exported from Upper Burma, then a British colony. The oil was transported in earthenware vessel to the riverbank, where it was poured into boat holds for shipping to Britain. In 1859 the US State of Pennsylvania became the unlikely birthplace of America's oil industry after the American industrialist Edwin L. Drake struck oil near Titusville. Initially producing around 10 barrels of oil per day, within two years the Titusville field was providing 3,000 barrels per day (480 m^3/d). The invention of oil refining led to the availability of kerosene as lamp oil, which has a clean combustion in contrast to then predominantly used whale oil. Due in part to overfishing, by the 1870s whale oil had become so expensive and hard to come that industrialists and merchants were forced to seek out alternative sources of fuel. This gap in the market was quickly filled with lamp oil, which would become known as "Pennsylvania kerosene." Break-bulk boats and barges were originally used to transport Pennsylvania oil in 40-US-gallon (150 litre) wooden barrels. But transport by barrel posed several problems. The first problem was weight: the standard empty barrel weighed 64 pounds (29 kilograms), representing 20% of the total weight of a full barrel. Also, the barrels tended to leak and could only be carried one way. Finally, the barrels themselves were expensive. For example, in the early years of the Russian oil industry, barrels accounted for as much as 50% of the cost of petroleum production.

DOI: 10.1201/9781003505044-1

Figure 1.1 Scottish engineer and pioneer, James Young.

DEVELOPMENT OF THE OIL INDUSTRY

The movement of oil in bulk was attempted in many places and in many different ways. Modern oil pipelines have existed in some form or other since at least 1860. The first oil tankers were two sail-driven tankers that were built in 1863 on the River Tyne in the north of England. The first ocean-going oil-tank steamer, the VADERLAND, was designed and built by the British shipbuilder, Palmers Shipbuilding and Iron Company, on behalf of the American-Belgian Red Star Line, in 1873. Despite her initial success, the vessel was soon curtailed by authorities who cited safety concerns. By 1871 the Pennsylvania oil fields were making limited use of oil-tank barges and cylindrical railroad tank cars similar to those in use today. In 1877, the sailing ship LINDESNÆS was converted to carry oil in bulk. The modern oil tanker was developed during the period between 1877 and 1885. In 1876, Ludvig and Robert Nobel, brothers of the Swedish chemist and philanthropist Alfred Nobel, founded Branobel (short for Brothers Nobel) in Baku, Azerbaijan. Throughout the course of the late 19th century, Branobel would become one of the largest oil companies in the world.

Ludvig Nobel was a pioneer in the development of early oil tankers. He first experimented with carrying oil in bulk on single-hulled barges. Turning his attention to self-propelled tankships, he faced a number of challenges. A primary concern was how to keep the cargo and fumes away from the engine room to avoid fires. The other challenges Nobel faced included allowing for the cargo to expand and contract in response to temperature changes and providing a method for ventilating the cargo tanks. In answer to these problems, Nobel signed a contract with the Swedish engineer and shipbuilder, Sven Alexander Almqvist in 1878, to design and build the world's first true type oil tanker. The result was the ZOROASTER, which sailed on its maiden voyage across the Caspian Sea from Baku, Azerbaijan, to Astrakhan, in southern Russia. The ship was built with Bessemer-process

Figure 1.2 VADERLAND.

forged steel, while the petroleum holds were iron. One tank was positioned forward of the midship engine room and the other aft of the engine room. The ship also featured a set of 21 vertical watertight compartments for extra buoyancy. For its time, the tanker ship ZOROASTER was technologically advanced, being reinforced with ballast tanks to enjoy better balancing during navigation during inclement weather. The ZOROASTER measured over 55 metres (180 feet) lengthwise with a breadth of over 11 metres (35 feet), for a draft of about 3.5 metres (10 feet). Around 240 gross tons (1,760 barrels) of crude and kerosene could be ferried by the vessel between the provinces of Astrakhan and Baku along the River Volga and the Caspian Sea route. In October 1878 Nobel ordered two more tankers of the same design: the BUDDHA and the NORDENSKJÖLD. By 1900, the Caspian Sea was crossed by 134 units built in the same way, for a total payload of 48,848 tonnage.

Another important landmark in the advancement of tanker technology was achieved in the early 1880s, when German-British capitalists and naval engineers met at the Armstrong-Whitworth Works in Newcastle upon Tyne. The first oil tanker with integrated hull tanks capable of crossing oceans with relative safety, anticipating many of the modern schemes to be incorporated into ship design, was designed in 1884 by the British naval engineer Henry Frederick Swan. Built mostly of steel with a number of innovative safety systems and an independent pumping station, the 2,700-ton vessel was the first ship in which oil could be pumped directly into the vessel's hull instead of being loaded into barrels or drums: the first ship ever to sail with "oil to her skin." The GLÜCKAUF was powered by a 992-horsepower steam propulsor, providing 11 knots at maximum speed. In addition, the GLÜCKAUF also carried sails for backup. At 318-foot long, 37-foot beam and 19-foot draft, the GLÜCKAUF was equipped with 14 tanks arranged in seven separate compartments each divided longitudinally by a continuous bulkhead. The pumping system and expansion boxes could unload the ship in as little as 12 hours. Laid down on 25 November 1885, the GLÜCKAUF was launched on 10 June 1886, sailing on her maiden voyage from the Tyne on 10 July 1886, arriving in Philadelphia with a cargo of 2,880 tons of crude. Built for Wilhelm Anton Riedemann's shipping firm in Geestemünde, the vessel mostly operated as a tramp steamer on charter to the Standard Oil Company. The GLÜCKAUF remained in service from 1886 to 25 March 1893 when she ran aground in heavy fog on Fire Island, New York. After the GLÜCKAUF was lost, Standard Oil purchased the GLÜCKAUF's sister ships.

In 1903, the Nobel brothers built two oil tankers which ran on internal combustion engines, as opposed to the older steam engines. The VANDAL, the first diesel–electric ship, was capable of carrying 750 long tons (760 tonnes) of refined oil and was powered by three 120 horsepower (89

Figure 1.3 ZOROASTER.

Figure 1.4 GLÜCKAUF underway.

kilowatt) diesel motors. The larger SARMAT employed four 180 horsepower (130 kW) engines. The first seagoing diesel-powered tanker, 4,500-ton MYSL, was built by Nobel's competitors in Kolomna, Russia. Nobel responded with the EMANUEL NOBEL and KARL HAGELIN, two 4,600-long-ton (4,700 tonnes) kerosene tankers with 1,200-horsepower (890 kilowatt) engines.

In 1902 the 475 feet (145 metres), the seven-masted schooner, THOMAS W. LAWSON, was built as the largest pure sail tanker. Designed to carry coal, and oil in barrels from Texas to the East Coast of the US, the 5,218 GRT schooner was later fitted out as an oil tanker in 1906. She was sunk in a storm off the Isles of Scilly, UK, on 14 December 1907, with the loss of 17 out of 19 crew, including the ship's pilot.

Figure 1.5 GLÜCKAUF grounded in heavy fog at Blue Point Beach on Fire Island.

Asian oil trade

The 1880s also saw the beginnings of the Asian oil trade to the East. The oil industry in Azerbaijan was the largest producer in the world at that time but was limited to the Russian market. To the west, John D. Rockefeller's Standard Oil Company dominated the world market. The idea that led to moving Russian oil to the Far East via the Suez Canal was the brainchild of two men: Marcus Samuel, and the shipowner/broker Fred Lane, who was the London-based agent for the De Rothschild Frères. Prior bids to move oil through the Suez Canal had been rejected by the Suez Canal Company as being too risky. However, Samuel approached the problem a different way. Instead of trying to force a ship through the Suez Canal, Samuel asked the company for the specifications of a tanker that it would allow through the canal. Armed with the canal company's specifications, Samuel tasked James Fortescue Flannery, ship designer for Bnito – the Caspian and Black Sea Oil Company, the Russian oil company of the Rothschilds – and ordered three tankers from William Gray & Company in Northern England. Named the MUREX, the CONCH and the CLAM, each had a capacity of 5,010 long tons of deadweight. In 1893 the Samuel brothers founded the Tank Syndicate together with Fred Lane and a consortium of Asian trading companies. In 1897 it was renamed the Shell Transport and Trading Company, forerunner of today's Royal Dutch Shell Company.

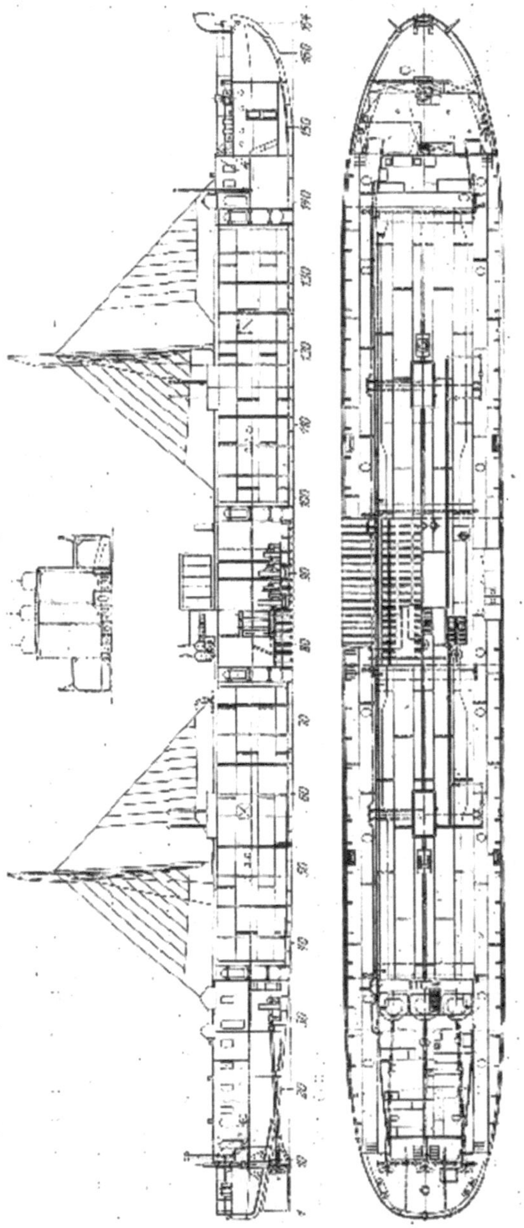

Figure 1.6 River tanker VANDAL (mechanical drawings), 1903.

Figure 1.7 THOMAS W. LAWSON on her maiden voyage, 1902.

With facilities prepared in Jakarta, Singapore, Bangkok, Saigon, Hong Kong, Shanghai and Kobe, the fledgling Shell company was ready to become Standard Oil's first challenger in the Asian market. On 24 August 1892, the MUREX became the first tanker to pass through the Suez Canal.

In the meanwhile, in 1890, the Koninklijke Nederlandsche Maatschappij tot Exploitatie van Petroleumbronnen in Nederlandsch-Indie (KNMEP) ("Royal Dutch Company for the Working of Petroleum Wells in the Dutch Indies") – part of Royal Dutch Petroleum – was founded. In 1892 the company struck oil near Pangkalan Brandan on Sumatra, just some months before Samuel's kerosene arrived in Singapore. At first, chartered ships were used, but in 1896 KNMEP launched its first tankers, the BESITANG and BERANDAN. The threat of the Tanker Syndicate was reduced as the Dutch government excluded them from trading in the Dutch East Indies. By the time Shell merged with Royal Dutch Petroleum in 1907, the company had 34 steam-driven oil tankers. Standard Oil started building tankers the same way as Shell and by 1900 was the owner of a fleet of around 60 tankers.

Pivoting again to the west, from 1912 the Compañía Mexicana de Petróleo El Aguila ("Mexican Eagle Petroleum Company"), founded in 1909 by Weetman Pearson to develop the newly found Mexican oil fields, and later

Figure 1.8 Royal Dutch Petroleum dock in the Dutch East Indies (now Indonesia).

Figure 1.9 Steam tanker CONCH, built in 1892 by W. Gray & Co. Ltd, West Hartlepool.

nationalised in 1938 as Pemex, also started to build its own tanker fleet. They quickly adopted Joshua Isherwood's new longitudinal framing system which allowed for the construction of much larger ships using a simpler construction process. Prior to the outbreak of World War I, the company owned a sizeable fleet of 20 tankers. Despite being the world's number one oil

Figure 1.10 FALLS OF CLYDE, the oldest surviving American tanker and the world's only surviving sail-driven oil tanker

producer, Standard Oil did not participate directly in the newly discovered oil fields of Texas and Oklahoma, which gave rise to opportunities for new oil companies to emerge such as Gulf Oil and the Texas Fuel Company, later renamed Texaco. Avoiding the use of Standard Oil's pipeline system, they started using tankers to get their oil to the East Coast. In combination with the oil fields discovered in Mexico and Venezuela, this caused a rise in the demand for tankers, which provided opportunities for the first independent shipowners to enter the tanker market, such as the Norwegian shipowner Wilhelm Wilhelmsen, who launched their first tanker in 1913.

World War I and the interbellum period

The fleet oiler USS MAUMEE, launched on 17 April 1915, pioneered the technique of replenishment at sea (RAS). A large ship for the time, with a capacity of 14,500 long tons of deadweight, the USS MAUMEE began refuelling destroyers en route to Britain at the outset of World War I. This technique enabled the US and Royal Navy to maintain their fleets at sea for extended periods, with a far greater range independent of the availability of a friendly port. This independence proved crucial to the Allies victory in World War II. Underway replenishment was quickly adopted by other navies. One example of this is the Australian fleet oiler HMAS KURUMBA which provided underway replenishment services to Royal Navy from 1917 to 1919. During World War I, unrestricted submarine warfare caused a shortage of tankers. So much so, in fact, that the US ambassador to the UK, Walter Hines Page, wrote

> the submarines are sinking freight ships faster than freight ships are being built by the whole world. In this way, too, then, the Germans are succeeding. Now if this goes on long enough, the Allies' game is up. For instance, they have lately sunk so many fuel oil ships, that this country may very soon be in a perilous condition – even the Grand Fleet may not have enough fuel.

The French president, Georges Clemenceau, also wrote of the problem in a letter to US President Woodrow Wilson, in which he stated:

> Gasoline is as vital as blood in the coming battles...a failure in the supply of gasoline would cause the immediate paralysis of our armies.

In response, Wilson directed the US War Shipping Board to commandeer all US-registered merchant vessels and also took charge of all American shipyards. An unprecedented budget of US$1.3 billion was approved by the US Congress. As part of this package of aid, the largest shipyard in the world was built on Hog Island, which later became synonymous for the Hog Islander.[1] Between 1916 and 1921, 316 tankers were built with a total capacity of 3.2 million long tons of deadweight; in comparison to the entire world fleet before World War I, which was just above 2 million tons, this feat of human engineering was quite miraculous. In 1923 about 800,000 long tons were laid up, which gave enormous opportunities for speculators, such as Daniel Keith Ludwig. In 1925 Ludwig bought the freighter PHOENIX, installing tanks into the holds. These tanks were riveted instead of welded, which often leaked, resulting in an explosive mixture in the PHOENIX's hull. Sadly, an explosion killed two crew members and severely injured Ludwig. Following this incident, Ludwig became a strong believer in welding. In 1928, the 16,436-ton CO STILLMAN was built by the German shipbuilder Bremer Vulkan. The CO STILLMAN was notable for being the world's largest oil tanker at that time, a record she held throughout her 14-year career.

For most of the 1920s, the major oil companies continued to develop their fleets by building increasing numbers of tankers; however, many independent shipowners also started to invest in tanker fleets of their own. By the outbreak of World War II in 1939, some 39% of the global tanker fleet was in private shipowner's hands. This was partly due to the legacy of the

1929 Wall Street Crash, where the vagaries of the stock market led to a sharp drop in worldwide economic activity resulting in mass unemployment and dropping freight rates for tankers. Despite this, the tanker industry has avoided the "Conferences," cooperation agreements that predominate in the scheduled liner trade. Tanker chartering is cited by economists as one of the very few examples of "perfect competition" – i.e., the price is set purely by negotiation between the buyer and seller of the service without regulation of outside bodies or rules. This explains why when supply of tankers exceeds demand, there is nothing to stop the price negotiated by the customers – the charterers – being driven right down to the floor. The early 1930s was such a time for tanker owners and adversity led them to explore new ways to cooperate. One such example was HT Schierwater, a resident of Liverpool whose own produce brokerage had fallen victim to the 1929 Crash. After his business failed, Schierwater joined the British tanker owner, United Molasses, which later developed into Athel Line, working first in its Liverpool office and then in London.

The Schierwater plan

Schierwater pioneered a plan to even the tanker supply-demand balance by cutting the number of tankers available to trade. The Schierwater Plan involved mothballing surplus tankers. The vessels that remained in trade would, from the expected higher freight returns, contribute 10% of their freight income to reward the owners of ships that had been mothballed. The Schierwater Plan was well received by the industry not least because the oil company charterers, concerned that the quality of the tanker stock should not fall due to underfunding, supported it and gave charter preference to ships entered into the scheme. The seeds of a cooperation between independent owners led to other opportunities to work together. On 28 February 1934, the International Tanker Owners' Association was inaugurated, with Schierwater appointed as its first Chairman to, in the words of shipowner Erling Næss:

> establish a medium for tanker owners to exchange information and opinions and to deal with the oversupply of tanker tonnage which depressed the market at the time.

The Schierwater Plan was the first such scheme of active cooperation amongst tanker owners to work. Although attempts were made to repeat its formula, no other scheme has enjoyed the same success. Whether its success was due to the participation of all parts of the industry, the forbearance of US anti-trust authorities or the recovery of economic activity and then the onset of World War II in 1939 is open to debate. Post-war, the massive reduction of tanker tonnage through sinking led to good times for tanker owners shipping oil from America and the Middle East to power the rebuilding of Europe. The Association was reassembled in 1949 by Schierwater and elected Reginald RS Cook of Hunting & Son, Newcastle, as its Chairman, a post he held until 1967. The Association continued to meet, but in times of prosperity, the meetings were more social than businesslike. Records show many involved dinners and golf matches between Britain and Norway. Despite the apparent success of the Schierwater Plan, there had been considerable doubts among some owners about its market effects. The continuing opposition resulted in the Association, which was the Plan's custodian, deciding in 1955 that a 76% vote of members in favour would be required to revive the mothballing plan.

In 1956, Egyptian President Gamal Nasser nationalised the Franco-British Suez Canal, leading to its closure and vastly increasing the length of tanker voyages from the Middle East oil wells to industrial Europe, and back again in ballast. Increased demand for tanker service brought good times but in 1957 the Canal opened again, now under Egyptian control, and the freight market again diminished. By 1962, after several years of poor freight rates, the Association was actively discussing a fresh mothballing plan modelled on Schierwater.

The international tanker recovery scheme

The international tanker recovery scheme was offered to owners and oil companies to improve conditions in the industry. The European oil majors British Petroleum (BP) and Shell were ready to back the scheme as the oil industry had backed Schierwater; however, this time the Americans were more reticent about joining the scheme. Nonetheless, the Association pressed on with the plan and at its peak in the summer of 1964 some 80 tankers (approximately 1.5 million tons of carrying capacity) were in lay-up under the scheme. However, the lay-up was short term and seasonal. By the winter of 1964/65, all the ships in lay-up left mothball and entered back into service. In 1967 the Suez Canal again closed only this time due to the Arab–Israeli War. Lay-up schemes were not the only objective of the Association. Erling Næss and Norwegian shipowner Jørgen Jahre promoted a Shipbuilders and Shipowner Collaboration Scheme in the 1960s to limit the construction of surplus tankers. European shipbuilders were ready to support the scheme; however, shipbuilders in the Far East, dominated by the Japanese, were too busy establishing themselves as the controlling force in world shipbuilding and were unwilling to back the scheme resulting in its failure to gather mass support.

Besides attempting to control the supply of tankers to manage surplus tonnage, the Association was active in attempts to harmonise safety requirements for tankers and to standardise equipment needs in different oil ports and terminals. This was before any international body for maritime affairs was established. The United Nations was only brought into being late in the 1940s. The Convention creating the Intergovernmental Maritime Consultative Organisation, the forerunner of the International Maritime Organisation (IMO), was drawn up in 1948 but not brought into effect until 1958/59. Shipping traffic was regulated by national rules and occasional bipartite treaties. The need for worldwide agreement on ship operation and equipment for a class of ship trading in a hazardous product was readily apparent to the Association. Following the stranding of the tanker TORREY CANYON on the British coast in 1967, and outcries about clean-up of oil spills and compensation for the victims of the maritime pollution, discussions started amongst tanker owners to form a voluntary agreement accepting liability and providing compensation funds, through insurance. The International Tanker Owners' Association was active in these discussions which led to the creation of the Tanker Owners' Voluntary Agreement concerning the Liability for Oil Pollution, or TOVALOP. The International Tanker Owners' Association was headquartered in London, signifying the continued strength of the British merchant fleet. Its membership was international, but the leading positions were those of the British tanker owners Hunting's of Newcastle, Athel Line, P&O and Erling Næss' Anglo Nordic. Stewart Browne of Athel Line and Vice-Chairman under Reginald Cook continued to serve as Vice-Chairman until the Association closed in London. Jan Hudig, Jørgen Jahre and Erling Næss would later become chairmen in

Figure 1.11 TORREY CANYON, 1967.

Oslo, Norway. The Association worked closely with the leading bodies of British shipping and with the London-based oil companies. Indeed, when in the 1960s the senior manager in BP's shipping operations, Houston Jackson, retired from BP, he was signed to Hunting's and elected Chairman of the Association in succession to Reginald Cook. Many of the British members were liner companies with tanker departments or sections closely monitored by the UK Chamber of Shipping. This led the more specialist tanker owners to become restless and concerned that other interests were limiting the effectiveness of the Association. By 1967, the closure of the Suez Canal had rendered the recovery scheme unnecessary. The Association's agenda had therefore shrunk, and the opportunity arose for independent tanker owners to recreate the Association on finer lines and with more independence from other influences. Although some of the Association's members resisted, the Association was wound up in London. Largely inspired by Jørgen Jahre, INTERTANKO, the International Association of Independent Tanker Owners, opened in Oslo at an inaugural meeting on 21 October 1970.

Closure of the Suez Canal

The oil spill caused by the TORREY CANYON in 1967 led to wider public awareness about the environmental dangers of oil tankers. In response, the major oil companies united in 1970 to form the Oil Companies International Marine Forum (OCIMF), which became instrumental in the drafting and implementation of MARPOL 73. In 1968, the International Tanker Owners Pollution Federation was founded to indemnify the victims of tanker incidents. For tanker owners, the Six-Day War of 1967 was of greater importance. In response to Israel's victory, the Suez Canal was closed to international shipping until 1975. This had the effect of sending freight rates skyrocketing (1) because of the shortage of tonnage, and (2) because ships had to round the African continent and pass the Cape of Good Hope. To offset the costs of longer passages, shipowners started thinking about designing and building larger capacity vessels. Given the limitations of the Suez Canal were no longer relevant, the only governing factor was how big is too big! In only a couple of years, the size of tankers quadrupled to more than 500,000 long tons and there were even plans for tankers of 1,000,000 long tons. In 1969 the first ultra-large crude carrier (ULCC) was built.

The world's largest super tanker ever built was constructed for Tung Chao Yung in 1979 at the Oppama Shipyard of Sumitomo Heavy Industries, Ltd. as the SEAWISE GIANT. This ship was built with a capacity of 564,763 DWT, a length overall of 458.45 metres (1,504.1 feet) and a draft of 24.611 metres (80.74 feet). She had 46 tanks, 31,541 square metres (339,500 square feet) of deck and was too large to pass through the English Channel. SEAWISE GIANT was renamed HAPPY GIANT in 1989, and then JAHRE VIKING in 1991. Between 1979 and 2004 she was owned by Loki Stream,

Figure 1.12 KNOCK NEVIS.

at which point she was bought by First Olsen Tankers, renamed KNOCK NEVIS and converted into a permanently moored storage tanker. Today, the Batillus class super tankers are the biggest ships ever constructed by gross tonnage.

Although the tanker fleet increased by around 12% annually in 1970, a shortage on tonnage remained. In 1973 this resulted in an enormous increase in new building orders, especially from oil majors that wanted to gain on the quicker deciding independents, who could ask enormous rates for their vessels. Where the existing tanker fleet comprised some 150 million long tons, in just a quarter of a year, tonnage of 75 million was ordered, although new build prices doubled. The increase in scale brought a new problem. Until then, the washing of tanks after cargo discharge was done by water. In December 1969 three tankers exploded during tank washing. The Dutch Shell tanker MARPESSA sank off the coast of Dakar and became the largest merchant vessel ever lost. The other two, the British Shell tanker MACTRA and the Norwegian KONG HAAKON VII sustained heavy damage but remained afloat. Shell investigated the incident and concluded that when water drops impact steel with high velocity, this generates static electricity that can cause explosions when combination with cargo vapours. This problem only became apparent with the large sizes of the tanks onboard very large crude carriers (VLCC).

Figure 1.13 MARPESSA.

Figure 1.14 MACTRA.

Figure 1.15 KONG HAAKON VII.

The solution was found by filling the cargo tanks with inert gas (IG), therein reducing the oxygen level such that the tank remains below the explosive limit of the cargo vapour. The use of IG is seen as the biggest step in increasing tanker safety. Ten years later however, 50 people were killed when the BETELGEUSE exploded at Whiddy Island in Bantry Bay. The Total tanker was still not fitted with IG. The vessel ENERGY CONCENTRATION did have this system, which prevented an explosion when it broke in two on 21 July 1980 during discharge at Europoort in the Netherlands. Washing with water in combination with a "Load on Top" system was replaced by crude oil washing (COW), a method developed by the British oil major BP. The advantages were cleaner tanks, no corrosive seawater in the cargo tanks and no discharges of polluted seawater overboard.

Era of the "super tanker"

Where the size of tankers had been more or less the same for 25 years, after World War II, they grew in size significantly, albeit initially slowly (Table 1.1). A typical T2 tanker of the World War II era was 162 metres (532 feet) long and had a maximum capacity of 16,500 DWT. In comparison, a modern ULCC can be as much as 400 metres (1,300 feet) long and have a capacity of 500,000 DWT. Several factors have encouraged

Table 1.1 Global tanker fleet 1957–1980

At year end	*Number of vessels in each tonnage class*			
	25–99 DWT	*100–149 DWT*	*150–199 DWT*	*>200 DWT*
1957	427	0	0	0
1958	568			
1959	715			
1960	826			
1961	892	2		
1962	989	4		
1963	1,092			
1964	1,226	6		
1965	1,303	15		
1966	1,395	34		
1967	1,446	59	5	2
1968	1,488	82	17	17
1969	1,535	96	30	61
1970	1,572	110	34	131
1971	1,600	125	37	200
1972	1,609	136	38	270
1973	1,656	150	41	357
1974	1,718	193	42	479
1975	1,714	241	47	588
1976	1,753	265	64	676
1977	1,580	279	76	712
1978	1,453	269	83	700
1979	1,435	304	45	699
1980	1,482	300	41	658

this growth. As discussed above, hostilities in the Middle East, which interrupted traffic through the Suez Canal, contributed, as did nationalisation of Middle East oil refineries throughout the 1960s and 1970s. Intense competition amongst shipowners also played its part. But apart from these considerations is the simple economic advantage that comes with larger vessels. The larger an oil tanker is, the more cheaply it can move crude oil, and the better it can help meet growing demands for crude products. It is worth noting that a similar evolution for container vessels has taken place since the mid-2000s. As of 2024, the largest container ships in service today are capable of stowing in excess of 23,000 20-foot equivalent ISO containers compared to the industry average of 5,000 in 2005. Whereas previously it was normal practice for the refinery of oil to take place near the well, this gradually moved towards the consumer location. Production in the Middle East developed, and the dominance of product tankers was replaced by crude oil carriers. Panamax tankers were built, soon followed by Aframax and Suezmax tankers.

After the war, it was expected that a large number of tankers would be laid up, which indeed happened. Despite the US Maritime Commission replacing the War Shipping Board, fraudulent activities are rampant. The Greek shipping magnates Aristotle Onassis and Stavros Niarchos used this time to buy tankers cheaply. The expected economic decline did not come, due in part to the Marshall Plan, with the demand for oil increasing to the point in 1947 that there was a shortage of tankers. Freight tariffs tripled overnight, enabling some to recoup their investment in as little as one voyage. Ludwig had started Universe Tankships in 1947 and began building larger tankers in his welding Shipyards. The BULKPETROL, boasting a massive 30,000 long tons, was the largest tanker to enter service at that time, though four of the five bulk class tankers sank, likely because welding technology was not yet fully understood. As larger ships could not be constructed in the yard at Norfolk, Virginia, Ludwig went to Japan where he introduced the concept of block construction at the Kure Naval Yard. Here, in 1952, the 38,000-long-ton PETROKURE was built. That same year, Onassis had a tanker of some 45,000 long tons built, with Niarchos following soon after. Both Onassis and Niarchos claimed to be the largest independent tanker owner in the world.

The SINCLAIR PETROLORE that Ludwig had built in 1955 was at 56,000 long tons, not only the largest freighter in the world, but also a self-unloading ore–oil carrier, the only one of that type ever built. It exploded on 6 December 1960 near Brazil. The likely cause being cargo leakage in the double bottom, which resulted in the largest spill until that time with an estimated 60,000 tons leaking into the South Atlantic. In 1956 the UNIVERSE LEADER of 85,000 long tons was built just before the Suez Crisis started with the seizure of the PANNEGIA. Within ten years, tanker sizes had quadrupled. In 1958, Ludwig broke the 100,000-long-ton barrier with the construction of the UNIVERSE APOLLO. This behemoth displaced 104,500 long tons, a 23% increase from the previous record holder, UNIVERSE LEADER. Not to be outdone, in 1962, Niarchos had the 106,000-long-ton SS MANHATTAN built. This was the largest merchant vessel ever built in the US. It was converted to have ice breaking capabilities in 1969 and was the first commercial ship to cross the Northwest Passage. Although the voyage was a success, a second attempt to cross the passage in winter proved impossible. Combined with numerous environmental concerns, the project was cancelled, and the Trans-Alaska Pipeline System built instead.

In 1966 the 206,000 long ton IDEMITSU MARU was the first VLCC to be built, followed shortly after in 1968 with the first ULCC, the UNIVERSE IRELAND. In as little as 20 years, the size of tankers had increased tenfold.

Oil crisis and consolidation

On 10 October 1973, the Yom Kippur War began, igniting the 1973 oil crisis, tripling oil prices to US$10 per barrel and halting economic growth. Newly build ships sometimes went straight from the yard to lay-up. The situation worsened when the Suez Canal was reopened in 1975. Just when the situation began to improve in 1979, the Iranian Revolution caused the second oil crisis, causing oil prices to rise to US$30. Ships were sometimes sent to the breakers after being in service for only ten years. It took until the end of the 1980s before many shipowners say any profits being made in oil transport. In 1979, World-Wide Shipping of Yue-Kong Pao, with a fleet of 204 vessels, many of them tankers, was the largest shipping company in the world with a tonnage of 20.5 million. Within five years, however, he had sold some 140 vessels. In 1980, Ludwig had built or acquired the largest fleet after Pao and CY Tung and was widely regarded the richest man in the US.

In 1976, the Intervention Convention was used for the first time when the US Coast Guard took on the salvage of the ARGO MERCHANT, although the vessel was in international waters. This was the first time the monopoly of the Flag State was challenged. The EXXON VALDEZ oil spill saw the introduction of legislation requiring tankers to have a double hull, a measure that was not universally seen as the best solution by all naval experts. Where a double hull should (in theory) minimise the consequences of a vessel grounding, Concordia Maritime developed an alternative approach with the Stena V-MAX, a VLCC fitted with two propellers, two rudders and two

Figure 1.16 During the salvage of the ARGO MERCHANT, the Intervention Convention was used for the first time.

Figure 1.17 STENA VICTORY, a V-MAX, approaching LOOP from anchorage.

redundant engine rooms. The key benefit being that a single fault would not result in a loss of steering, therein reducing the chances of grounding. That being said, the sheer size of VLCCs and ULCCs limits their sailing area and available ports. In the US, Louisiana Offshore Oil Port (LOOP) is the only facility that is able to handle VLCCs. To overcome this, ULCCs and VLCCs often anchor offshore before lightering to smaller tankers that are able to reach the destination port. The largest oil terminal in operation (as of 2024) is the Ras Tanura Oil Terminal in Saudi Arabia.

Emergence of the ultra-large crude carrier (ULCC)

In 1999 the Greek Hellespont Steamship Corporation ordered four double-hulled super tankers to be built between 2002 and 2003. These sister ships, the HELLESPONT ALHAMBRA, HELLESPONT METROPOLIS, HELLESPONT TARA and FAIRFAX, were sold to the Overseas Shipholding Group and Euronav in 2004. Renamed the TI class, and currently registered as the TI ASIA, TI EUROPE, TI OCEANIA and TI AFRICA, are as of 2024 the world's four largest working super tankers. Each of the four sister ships has a capacity of over 441,500 DWT, a length overall of 380.0 metres

Figure 1.18 EXXON VALDEZ.

(1,246.7 feet) and a cargo capacity of 3,166,353 barrels (503,409,900 litres). The first ULCC tankers to be built for some 25 years, they were also the first ULCCs to be double-hulled. To differentiate them from smaller ULCCs, these ships are sometimes given the V-Plus size designation. In February 2008, the owners of the TI AFRICA and TI ASIA began converting both vessels into stationary floating storage and offloading (FSO) units. The TI AFRICA and TI ASIA were anchored at the Al Shaheen Oil Field near Qatar in 2009.

With the exception of the pipeline, the tanker is the most cost-effective way to move oil today. Worldwide, tankers carry some 2 billion barrels (3.2×1,011 litres) of crude oil each year, with an average residual cost of only US$0.02 per gallon at the pump.

DEVELOPMENT OF THE CHEMICAL AND PRODUCT TANKER

In the years following World War II, a series of chemical industries sprang up along the US Gulf Coast. These new industries relied on Texas oil and gas fields and Louisiana sulphur mines to provide the raw chemical feedstocks. Initial plant production figures were insignificant compared with today's mammoth outputs, thus enabling shipments to be made to consumers

Figure 1.19 HELLESPONT ALHAMBRA (now TI ASIA).

along the Atlantic Coast in drums, portable tanks and railroad tank cars. Throughout the 1950s, however, demand for chemicals quickly increased and more extensive and sophisticated means of transport were required. For a while dry cargo ships with deep tanks were able to supplement existing methods of transport but the appearance of hazardous new chemicals which had to be shipped in large batches made it apparent that a new type of sea-going vessel was required. The first chemical carriers were converted war-built T2 tankers. By realising the significance of cargo segregation, the tank layouts in the earliest of these conversions enabled the simultaneous carriage of several hazardous and incompatible cargoes. The first of the new breed was the 9,073 gross tonnes RE WILSON, which was converted for the Union Carbide and Carbon Corporation in 1948. The RE WILSON was fitted with a double bottom and deep well pumps, unique for such ships at that time. Her centre tanks allowed the carriage of nine different chemicals, while petroleum products of moderate density, such as kerosene, could be carried in the ship's wing tanks. She entered service in January 1949 and shuttled regularly from the Gulf Coast ports to New York, except for a period of about 18 months in the mid-1950s when she carried chemicals from Texas City to Los Angeles and San Francisco. The RE WILSON operated successfully until 1971 when she was scrapped in Spain. Not all the US' chemical tankers began life as T2 tankers though; one such vessel was the Texan. Built in 1946 as a C4 cargo vessel, she was converted and lengthened in Japan in 1954 to an ore/oil vessel and in 1957 the ore holds were converted to carry 14,000 tonnes of chemicals. These vessels operated regularly between the Gulf and West Coasts until January 1975 taking chemicals out to the West Coast and returning home with between 7,000 and 8,000 tonnes of lube oil and other clean petroleum products. This was the genesis of the parcel tanker trade. Parcel tankers are smaller sized vessels which carry small lots of liquid chemicals or "parcels," usually on a regular service. Parcels may be anywhere from a few hundred to a few thousand tonnes each; they can be of any of a multitude of products; and they could be loaded and or discharged at any one of several ports along an established route.

The earliest parcel tankers, like early chemical tankers, were conversions based on petroleum product tankers built in the late 1940s and early 1950s. These ships had been laid up after the post-Suez collapse in 1957, as they were deemed uncompetitive compared to the larger, newer vessels. As the international trade in chemicals was developing rapidly; however, shipowners realised the potential of having smaller volume tankers and were willing to invest money on a small amount of conversion work to prepare these vessels for time charters to keep them employed. Conversion work usually entailed adding a few bulkheads to provide smaller tanks, coating some of the tanks with zinc silicate, installing additional pumps and pipelines to provide segregation and, if necessary, adding a second pump room. At the time daily running costs were low (between US $2,000 and US $3,000 were common for 12–18,000 DWT parcel tankers), enabling low time charter rates (as low as $3.00 per DWT per month). It was these low freight rates combined with efficient handling of difficult and hazardous products which gave the parcel tanker trade its initial boost. In the early 1960s typical rates were averaged around US $8.00–$10.00 per tonne on the US Gulf to Rotterdam route and between $14 and $18 per tonne on the US Gulf to Japan route. There is little doubt that the artificially low rates provided by the worldwide parcel tanker services had a catalytic effect on the growth of the chemical industry. These low freight rates, however, have at times come back to haunt the parcel tanker operation as they have never truly reflected the actual cost of ship construction or replacement. At the time, the charter rates that were paid for these ships as parcel tankers did not include the amortisation of any capital element. As a result, some operators requiring replacement tonnage in later years were hard pressed for capital funding. As the first purpose-built parcel tankers appeared in the early 1960s shipbuilding prices were still comparatively low. At this time, the European petrochemical industry was slowly recovering from the war, and so it was left to US chemical manufacturers to supply the bulk of European customers. In addition, the world trade in edible oils, lubricating oils and inorganic chemicals was expanding. This led to several operators deciding to get enter the parcel tanker sector. The first purpose-built ships incorporated the main characteristics of the early converted parcel tankers in addition to several novel technologies. More bulkheads were included in the cargo spaces to provide the ship with upwards of 40 tanks. Many ships incorporated a variety of coatings on a single vessel to ensure compatibility with a wider range of cargoes. Stainless steel tanks, to enable vessels to carry corrosive cargoes requiring a high degree of product purity, were fitted to many vessels. Other features included installing heating coils or ducts and sophisticated safety, alarm and IG systems.

The types of products that parcel tankers transport in bulk, i.e., chemicals, edible oils, lubricating oils and solvents, necessitated trade patterns. In general terms, chemicals produced in the US, Japan and Europe, which incidentally are also the main market for these chemicals, are also required in South America, South Africa, Australia and the Indian Ocean region. Therefore, only the difference between output and demand is moved by sea between individual countries. Since the parcel tanker acts as a buffer to balance the petrochemical plants output programme, no long-term consistency can be expected regarding the movement of any one product. Edible oils are produced either in the agricultural areas of the Western Hemisphere or in the tropical areas of the Far East and Africa. Consumption is centred in developed countries which do not produce the raw materials necessary for making soap, cooking fats, margarine, etc. Hence, there is a consistent flow of products such as soya bean oil, palm oil, beef and mutton, tallow, coconut oil and so forth, from the producing areas into Europe, the US and Japan. Lubricating oils and solvents are manufactured in the refining centres of the world and are widely distributed to countries which may have a low consumption which would not warrant local manufacture. Patterns of loading enormous quantities at one port and discharging small quantities at several remote ports are common. Unlike chemicals, a consistent pattern can be seen with lubricant and solvent movements. Parcel tankers are influenced by global trends more than any other sector of shipping. The closure of the Suez Canal in 1967 with its doubling of the voyage distance from the Persian Gulf quickly drove rates upwards. Parcel tanker operating costs, which had increased at about 6% annually during the early 1960s, jumped 50% between 1969 and 1971 and have consistently climbed at almost 12% per year since then. In addition, many operators switched to the lucrative long-distance clean and dirty petroleum products trades.

The IMO's requirements for double-bottomed, double-skinned vessels, individual cargo tank venting, containment of slops and ballast and approved stowage of an extensive list of commodities have further pushed up the construction and operating costs for parcel tankers. As a result of price increases, the price of bunkers has escalated and has become a major part of the operating cost. During 1974 the freight rates for parcel tankers began to reflect these cost increases. Although the tanker market collapsed in the first few months of that year the parcel market stayed firm with rates finally covering ship replacement, i.e., US Gulf to Rotterdam rates reached US $75 per tonne and US Gulf to Japan rates increased to US $150 per tonne. By late 1974, however, the worldwide recession resulting from the increased oil prices finally hit the international chemical market, while fats and oils formed unusual new trade patterns. During the period 1974–1978 freight rates remained low causing operators major financial problems. This was especially felt by those companies that had just commenced large building programmes. In 1978 the MARPOL Code was fully implemented having been ratified in 1972. This meant that shipowners who had not used the preceding six years to upgrade their pre-1972-built ships would be heavily penalised. In the last quarter of 1978, an unexpected uptick in the parcel tanker sector returned. From September 1978 to early 1979, rates soared, reaching two to three times those of six months earlier. Those companies that could engage in the spot trade operated very profitably, while other operators with a year's contract of affreightment only enjoyed the tail end of the boom. That said, the boom was short lived. The world recession, together with continuing overcapacity, returned the sector to low rates and subsequently low revenues. Since the brief period of reasonable freight rates that occurred between 1978 and 1979, the parcel tanker owner has had precious little to be optimistic about, with overheated competition and surplus capacity. During this period substantial upgrading of the parcel tanker fleet meant that vessels underwent significant modernisation through the adoption of increasingly sophisticated technology. This, of course, was extremely costly. The late 1983 price for a 35,000 DWT parcel tanker capable of more than 50 cargo segregations was in the region of US $40 million in the Far East, and more in Europe. At the MariChem Conference in Hamburg in 1983, it was universally accepted that parcel tanker operators would need better returns if they were to remain competitive.

To qualify this assertion, the example of a 30,000-DWT parcel tanker was used operating on a transatlantic route between Rotterdam, New Orleans, Houston, Texas City, before returning to Rotterdam via Antwerp. With 13 days of port time, and 30.3 days at sea, the entire round journey would take an average of 43.3 days. With an average daily cost of US $25,500 at sea and US $20,650 in port, the total cost of the voyage, including port expenses, was calculated at US $1,121,100. With a westbound ballast leg and a 100% cargo load coast bound, the vessel would require an average freight rate of more than US $38 per tonne to merely break even. A cursory look at the figures above demonstrates the actual rates paid throughout 1983 were well below this figure. Demonstrating the way that costs have steeply rising, the Conference noted that a similar sized ship of an equivalent age on such a voyage in 1976 would have required a freight rate of only US $27 per tonnes to cover its operating costs. Stanelift's conclusion was quite stark: the prognosis for the parcel tanker trade was not good and would become significantly worse if finance houses such as banks and capital investors refuse to invest in these types of ships. As such, it was widely recognised that the smaller parcels, demand for more sophisticated conditions of containment for cargoes of immense value or hazard, and increased requirements for stainless steel or special coatings would certainly push operating and construction costs for parcel tankers even higher. In addition, any increase in ship size is negated by longer stays in port as many terminals have only limited reception facilities. Moreover, it is frequently the case that a ship must shift several times during a single loading and discharging operation. To compound matters further, it is often the case that large parcel tankers spend as much as one-third of their time in port.

The lack of profit inherent in parcel tanker operations has led an increasing number of operators to become involved in both storage and terminal operations, and with inland transport. In the 1980s the Norwegian companies Odfell Westfal-Larsen, for instance, operated five specialist terminals throughout the world, including the Bay tank facility in Houston, Texas and Stolt-Nielsen also increased its terminal interests, whilst operating in excess 700 specialist containers and several inland craft on the US waterways. This has led to an increase in rationalisation and consolidation within the parcel tanker sector, with fewer but larger operators, of which Odfjell and Stolt-Nielsen are the largest.

Figure 1.20 STOLT HAGI chemical/parcel tanker.

Note

1 Hog Islanders is the slang for ships built to Emergency Fleet Corporation designs numbers 1022 and 1024. These vessels were cargo and troop transport ships, respectively, built under government direction and subsidy to address a shortage of ships in the United States Merchant Marine during World War I. American International Shipbuilding, subsidised by the United States Shipping Board, built an emergency shipyard on Hog Island at the site of the present-day Philadelphia International Airport. Although no ships were built in time to participate in World War I, many ships were active in World War II, with roughly half of those produced at Hog Island being sunk during that conflict.

Chapter 2

Typical tanker arrangements

GENERAL

Oil tankers (Figures 2.1–2.5) can be divided into three core sections: the fore part, the tank area and the after part. The reasoning behind this design is that means must be provided to keep potential spills away from the crew accommodation and navigation areas of the vessel. Segregated ballast tanks (SBTs) help contribute towards the protection of the marine environment. There are requirements with respect to the number and minimum capacity of slop tanks, the cargo system, and the stripping system. Whilst they are widely used, it is worth noting not all tankers have separate stripping systems. For the majority of tankers, steam heating coils are used for heating the cargo tanks, which helps keep the cargo viscous. A large proportion of this trade consists of the transportation of crude oil, but refined products may also be carried in considerable quantities and include fuel oil, diesel oil, gas oil, kerosene, gasoline and lubricating oils. The design of a tanker must consider the particular trade for which it is intended. A high rate of loading and discharging is desirable; pumping capacity and size of pipelines are important in this respect. The safety factor must be borne in mind with the provision of a fire-smothering installation and the provision of cofferdams at the ends of cargo spaces, ventilating pipes to tanks, etc. Ships intended for the carriage of heavy oils would have steam heating coils fitted in tanks. The cargo space is divided into three sections athwartships by means of two longitudinal bulkheads and individual tanks by transverse bulkheads. The maximum length of an oil tank is 20%L (L is the length of the vessel) and there is at least one wash bulkhead if the length of the tank exceeds 10%L or 15 metres. Tanks are numbered from forward, each number having port, centre and starboard compartments. Pump rooms are often located aft so that power may easily be supplied to the pumps from the engine room, but ships designed to carry many grades of oil at once may be fitted with two pump rooms placed so as to divide the cargo space into three sections. The system of pipelines used in a tanker is such that great flexibility is possible in the method of loading or discharging, and different parcels of cargo may

DOI: 10.1201/9781003505044-2

Figure 2.1 Ore-bulk-oil carrier, SKS TANA.

Figure 2.2 Deck view of the ore-bulk-oil carrier, MAYA, 2006.

be completely isolated from one another during loading and subsequently during discharge. In some cases, a small, separate line is used for stripping the last few inches of oil from each tank.

These vessels were provided with a midship pumproom which contained two steam reciprocating pumps for handling cargo. To control the flow of liquid when the vessel was rolling in a seaway, and to avoid large areas of free surface, the tanks were provided with trunk ways, which reduced the area at the top of the tank. Vessels, however, were often far short of their marks when loading light products, later types began to incorporate the "summer tank" which was housed on the trunk deck and was filled by means of a drop valve from the main tank below (as shown in Figs.2-1 and 2-2).

Towards the middle of the 1920s, the twin bulkhead ship made its appearance. Slowly but surely the advantages of the contemporary design made itself felt, and the centre line bulkhead type began to be replaced in all but a few special types and coasters, where size made the twin bulkheads impracticable. Welding was used in ship construction for a considerable period before World War II. However, where hull construction was concerned, welding was always viewed with grave suspicion, but like all new methods, materials and techniques improved, and during World War II whole ships were constructed on this basis. The advantages of welded hulls are obvious: all the plates are welded in a straight line, and there are no plate landings to restrict the flow of water along the hull as the vessel is propelled through the water. In addition to this, rivets tend to work; leaks from this source are quite frequent both in the hull and in the bulkheads separating the cargo tanks. Welding has eliminated leakage of this nature. Over the last 20 years, a great deal has been learnt about the use of metal in all types of construction. Research into metal fatigue and wastage as well as the use of coatings to prevent this, has helped to simplify some of the problems encountered when carrying highly corrosive hydrocarbon liquids. Large-scale models in ship model basins have assisted ship designers in examining stress problems and simplifying the design and layout of large tankers, thus reducing the cost of construction.

Where once a large crude tanker could be expected to have a minimum of 36 cargo compartments and a whole battery of pumps, pipelines and valves, the modern tendency is to reduce the number of tanks and other equipment so that a ship of 200,000 tonnes deadweight tonnage (DWT) built to American Bureau of Shipping (ABS) or Lloyd's specifications may only have 15 cargo compartments: with individual tanks holdings as much as 140,000 bbl. or 20,000 tonnes of oil. The tendency is also to reduce the number of pumps and to install fewer and more powerful units with a higher capacity head. In some cases, suction piping has been eliminated by the introduction of suction pipe ducts and or sluice valves. The extensive use of sluice valves has led to the implementation of the *Free Flow System*, which has come into widespread use where suction piping has been eliminated. Such systems have definite advantages, particularly in capital savings when the ship is built. There are, however, definite disadvantages from an operational perspective, namely when more than one grade of oil is carried and when during tank cleaning and ballast changing operations. Apart from the layout of the cargo compartments and pumping systems, there have been significant changes in other directions, for example, with power-operated valves and remote controls becoming increasingly common. Safely used and maintained, such improvements show an economic return by reducing manpower requirements and eliminating human error from a complex operation where expensive equipment can be seriously damaged. It would not be wise to neglect other areas where considerable changes have taken place. Modern tankers are designed with no accommodation amidships. The bridge and living accommodation are exclusively located aft. Safety and economics have been the main reasons for this change and the arguments of masters and pilots who have opposed it on navigational and ship handling grounds have enjoyed precious little support today. In 1974 tankers were classified by size for freight purposes:

(1)	General purpose vessels	15,500/24,999 DWT
(2)	Medium range	25,000/49,999 DWT
(3)	LR1 (Large Range 1)	45,000/79,999 DWT
(4)	LR2 (Large Range 2)	80,000/159,999 DWT
(5)	VLCCs or very large crude carriers	160,000/320,000 DWT
(6)	ULCCs or ultra-large crude carriers	320,000 DWT and above

While very large crude carriers (VLCCs) and ultra-large crude carriers (ULCCs) were, and always are, likely to be exclusively engaged in the carrying of crude oil, handy size medium size vessels tend to cover a wider range of duties.

The larger vessels in the Large Range 2 (LR2) range – i.e., over 100,000 DWT tend to be crude carriers. They trade between ports that are restricted by draft or other limitations so that VLCC or ULCCs cannot be used.

Large Range 1 (LR1) and LR2 range vessels of less than 100,000 DWT are divided into two classes: (1) dirty product carriers; and (2) clean product carriers. The larger dirty product carriers are frequently switched between the crude trade and carrying refined dirty products. After carrying crude, the cargo tanks must undergo cleaning to remove wax and crude residues which might affect the flash point of dirty products such as fuel oil.

Figure 2.3 Typical oil tanker, 1950.

Some large dirty product carriers have their tanks coated to reduce corrosion from crude and water washing, and to facilitate changes of one cargo type to the other. Clean product carriers in the medium-size range tend to be less than 50,000 DWT. Many are purpose-built with coated tanks and have sophisticated pumping systems capable of handling 12 or more grades. The largest dirty and clean product tankers have evolved from changing trade practices and requirements and though some of these vessels may be involved in short-haul coastal distribution of refined products many are now involved in longer haul work.

General purpose tankers cover the largest range and variety of different cargoes carried. This size range includes some chemical carriers as well as a host of purpose-built clean and dirty product carriers engaged in short-haul and coastal distribution. Tankers smaller than 16,500 tonnes DWT are clean or dirty product short-haul coastal vessels, but some are built to handle special products like bitumen, chemicals acids as well as lubricating

Figure 2.4 Typical crude oil tanker, AMYNTAS, St. Nazaire, France, 2016.

oil. The big building programmes of the 1960s and early 1970s were the result of high freight rates. The large numbers of ULCCs and VLCCs which came into service received a lot of publicity and to some extent hid the fact that the number of smaller ships produced was also significant. The 1973 oil price increase was a major catastrophe which reduced oil consumption around the world and made many ships of all sizes and classes redundant. The tanker building boom had produced a spate of larger ships. Quite a number were over 500,000 DWT. Four of these vessels operated under the French flag: two were owned by the Société Maritime Shell and two by the Compagnie National de Navigation. All four had similar dimensions: the overall length was 1,359 feet and the beam was over 200 feet. Each ship was driven by steam turbines and twin screws. The carrying capacity of all four differed slightly but was in the region of 550,000 DWT. All four ships had short lives under the French flag and were laid up and eventually sold for scrap. The largest tanker built at that time was the SEA WISE GIANT, which was owned by the Island Navigation Company. She was originally built as the OPPIDA before undergoing an enlargement programme in Japan. She had the highest recorded deadweight of 564,739 tonnes. In 1988 the vessel was reported to be on fire and suffered severe damage after a bomb attack

Figure 2.5 Oil products tanker CELSIUS RICHMOND at BP Oil Refinery Jetty, Kwinana, December 2021.

in the Persian Gulf. Whilst the layup and scrapping of ULCCs and VLCCs received significant publicity, all classes and sizes of ships were negatively affected by the reduction in the consumption of oil. Even chemical tankers, which might have expected to escape the worst, were similarly affected.

To understand the problem, we must look beyond the immediate effect of the OPEC price rise. During the 1960s studies show that as an energy source, oil consumption grew faster than any other fuel source. At over 60% in terms of growth, oil was by far the world's most popular fuel. As a result, the oil tanker increased in size and numbers. In the 1970s oil was still immensely popular despite price increases. As a percentage share of growth in world energy oil grew by 44%, but it must be said that most of the recorded growth occurred in the first half of the decade. During the first half of the 1980s, the growth of world oil consumption was less than 10% but improved with the decline in oil prices in the second half of the decade. The oil tanker and freight market are dependent on oil consumption. It has become apparent that the 1973 price increase, which triggered the global reaction, drastically reduced oil consumption. Apart from oil tankers laid up and scrapped, many refineries were shut down and some were even dismantled. It became clear that many oil companies saw the reduced consumption of oil as a long-term terminal decline, rather than a short-term phenomenon. Oil company fleets were drastically reduced in size, often by selling individual ships to independent owners using charter back agreements as bait. When this failed even modern ships were laid up and scrapped. Furthermore, the seven-year war between Iran and Iraq was responsible for the withdrawal of a significant volume of tankers from layup. Many of these vessels were severely damaged with others lost completely. The war between Iran and Iraq had significant repercussions for the oil industry worldwide. Insurance premiums for trading in war zones reached astronomic proportions. Towards the end of 1988 crude oil prices started to increase and by early 1989 had reached US $19 a barrel with freight rates responding accordingly as the demand for oil picked up. In turn, this resulted in a steady reduction in the number of laid-up tankers. The crisis caused by Iraq invading Kuwait in August 1990 also caused prices to rise significantly. Crude oil on the open market started selling at US $30 a barrel with some authorities considering US $40 a barrel a very real possibility (it is worth mentioning that as of July 2022, the average price per barrel is trading at US $99). The effects on oil transportation of such a crisis are hard to predict in terms of tanker demand. One trend that emerged throughout the 1990s is that whilst larger industrial nations maintained their reserves by topping up with imports there was an increasing tendency for oil companies to buy and load crude oil and use tankers as offshore storage against further price rises.

TYPES OF TANKERS USED FOR THE CARRIAGE OF OIL

Although the term "oil tanker" is widely used to denote any type of vessel capable of transporting oil, there are in fact several classes of vessels which are used for this purpose. *Crude oil tankers* are vessels designed and built to exclusively carry crude oil. *Product tankers* are ships engaged in the trade of carrying oil other than crude oil. This includes refined oil, lubricating oil, gas oil, and so forth. *Combination tankers* are the first of three unusual types of tankers in that they can carry both oil (as a liquid) and solid cargo in bulk. The second type of unusual tanker is the *oil/bulk/ore carrier* or OBO. Similar in design to the combination carrier, OBOs have a centre compartment which is used for liquid or solid cargo, but unlike the combination carrier, the wing tanks are used exclusively for ballast. The third type is the *product/bulk/ore carrier* or PROBO. Similar again to the OBO, PROBOs have a more sophisticated design which allows them to carry higher specification liquid petroleum cargoes and some liquid chemicals such as caustic soda. It is worth noting that although OBOs and PROBOs were common throughout the 1970s and 1980s, by the mid-1990s both ship types had fallen out of popularity due to their high construction and maintenance costs, but also due to the significant amount of training the crew needed to undertake given the disparity in cargo behaviour and management.

RULES AND REGULATIONS

Safety rules are critical to keeping a ship insured. Most safety rules with respect to shipping originate from the International Maritime Organisation (IMO), and come in the form of:

- International Conventions[1]
- Codes of Construction, and
- Codes of Safe Practices.[2]

In addition to these "international" rules, national rules are equally important as ships must be built and operated in accordance with national laws and Flag State requirements. For instance, vessels built in the UK must comply with Maritime and Coastguard Agency, UK (MCA)-mandated safety standards, even if the vessel will sail under a Panama flag. The vessel must then comply with Panamanian Flag State standards. Furthermore, shipowners and operators must comply with Classification Society[3] (Class) rules. These are important for maintaining the vessel in a safe and seaworthy condition. Class rule standards are maintained through periodic surveys and certification regimes. Though independent of government, Classification Societies are authorised by Flag State authorities to oversee the implementation of vessel standards. To ensure these standards are adhered to, Flag State authorities conduct Port State inspections. Any deficiencies in standards can result in penalties and in the worst cases, the arrest and impound of the vessel and crew.

IMO Conventions that govern the safe operation of all ship types

All ships, including tankers, must comply with the:

- International Convention for the Safety of Life at Sea (SOLAS) 1974, as amended,
- International Convention for the Prevention of Pollution from Ships, 1973, as modified by the 1978 Protocol (MARPOL73/78), as amended, and
- International Convention for the Standards of Training Certification and Watchkeeping for Seafarers (STCW, 1978), as amended.

For ships operating in US waters, they must also comply with the Oil Pollution Act, 1990.

Bulk chemical codes

Bulk chemical carriers must, in addition to those listed above, also comply with the:

- International Code for the Construction and Equipment of Ships Carrying Dangerous Chemicals in Bulk (International Bulk Chemical Code – IBC Code), and
- Code for the Construction and Equipment of Ships Carrying Dangerous Chemicals in Bulk (BCH), and
- Standards for Procedures and Arrangements Manual (PA Standards)

Bulk gas codes

Ships engaged in the transportation of bulk gas cargoes must comply with the:

- International Code for the Construction and Equipment of Ships Carrying Liquefied Gases in Bulk (IGC Code), and
- Code for the Construction and Equipment of Ships Carrying Liquefied Gases in Bulk (GC).

TYPES OF PIPELINE SYSTEMS

Pipeline systems on tankers differ in their degree of sophistication, depending on the employment of the tanker. ULCCs and VLCCs have simple pipeline systems, typically consisting of a direct line system whereas some product (parcel) tankers may have very sophisticated piping systems. This could be the ring main system or in the case of chemical product tankers it may mean an individual pipeline and an individual pump for every tank on board. There are three systems of pipelines found on tankers and a fourth system that is used on large crude carriers. These are the:

- Ring main system,
- Direct line system,
- Single line to single tank system (chemical and product ships), and
- Free flow system (crude tankers).

Ring main system

The ring main system is of a square or circular layout. It is used mostly on product tankers, as the full segregation of cargo is required. The system is expensive as it uses more piping and has extra numbers of valves. However, if the vessel is carrying many grades of cargo, the advantages compensate for the extra cost of the original outlay.

Direct line system

This system is found on crude oil carriers where up to three grades of cargo are carried as most of the direct pipeline systems are fitted with three direct lines. This system is cheaper to construct. The main disadvantage of the ring main system is that line washing is more difficult to conduct, the system has fewer valves which makes pipeline leaks difficult to control, and the system lacks versatility through reduced valve segregation. This system allows the vessel to carry as many grades of oil as there are tanks. The general disadvantage of the direct line system is the cost factor of having a multitude of pumps onboard.

Free flow tanker

This system is usually found on large crude carriers, where the cargo piping is not used for the discharge of cargo. Instead, gate valves are provided on the bulkheads of the tanks which when opened allow the oil to flow freely into the aftmost tank and into the cargo control room (COP). The advantages of this system are primarily the cost factor as it allows for faster drainage and efficient means of pumping the cargo tanks. The main disadvantage is that only crude can be carried.

Independent system

This layout is not common in the tanker trade though is quite normal on chemical ships. There are some product tankers that also have this system installed. The system consists of a single line servicing an individual tank through an independent pump that could be either a submersible or a deep well pump.

Lines and piping

Bottom lines

Here, the vessel is fitted with four centre tanks and five pairs of wing tanks for cargo. The cargo main lines are in the vessel's centre tanks. With the term "bottom lines" we understand that the location of these lines will be on the bottom of the vessel, usually supported about four to six feet above the vessel's hull. Crossover valves, i.e., two valves on each crossover, connect the bottom lines to each other. When carrying more than one grade, a two-valve segregation is used to ensure compliance with the regulations. From the drawing we can establish that, from the bottom lines, there are lines, which lead to each cargo tank. These lines end on the cargo tank suction bell mouth. Each bottom line serves its own set of cargo tanks. For example, bottom line no. 1 serves CT1 and WT5 P/S. Bottom line no. 2 serves WT1 P/S and CT4. Bottom line no. 3 serves WT2 P/S, CT3 and WT6 P/S.

Drop lines

From the manifold area on the main tank deck, the drop line is connected to the deck mainlines which leads to the bottom lines. In the drawing below we can see the drop line and the drop valves on the lines leading vertically downwards from the main deck lines to the cargo lines in the vessel's hull. These drop lines are used during loading. By closing the deck line's master valves, the cargo is led to the vessel's cargo tanks when using these drop lines. This means the pump room is completely isolated from the cargo during loading. However, during discharging the drop lines are isolated from the cargo by keeping the drop valves closed. Despite this, it is imperative to maintain a close watch on the pump room for any signs of leaks.[4]

Pump room piping

On crude tankers, the pump room is the main point between the cargo tanks and the main deck, all the way to the manifold, where the ship lines are connected to the shorelines. From the cargo, tank the bottom lines lead all the way to the main cargo pumps. To simplify things, we can divide the pump room into two parts. The first part is called the cargo pump free flow side; the second part is called the cargo pump delivery side. These sides are commonly referred to as the *suction side* and the *pressure side*. It is worth noting that centrifugal pumps do not have any suction ability. On the cargo pump free flow side, the bottom lines end at the cargo pumps. On this side, some crossover lines connect the systems to each other. The first crossover after the tank area is the stripping cross, marked on the drawing as "crude oil suction -x-over line." The stripping cross is located crosswise from the bottom lines and connected to the bottom lines with pipe bending and valves. By using this crossover, it is possible to discharge from cargo tanks online system no. 2 with COP no. 3 and so on. Further towards the COP, on the bottom lines, there is a valve on each of these lines, usually called the "bulkhead valve." This is because the location is normally close to the bulkhead, separating the cargo tank area and the pump room area. Further on the free flow side of the cargo pump, is the seawater suction crossover line. This line is also crosswise from the bottom lines and is connected to the sea chest on each side (port and starboard). This line supplies the cargo pumps with seawater during the water washing of the tanks and lines and is used when ballasting for departure if necessary. Crossing between different lines and pumps is also a possibility with this crossover line.

Leaving the free flow side of the system, the next stage is the delivery side of the pumps. The first stop is the first valve after the cargo pumps, the delivery valve or throttling valves. They may also be referred to as discharging valves or pressure valves though the most descriptive is "delivery valve." With this valve, we can adjust the backpressure and the load conditions for what the pump is going to work against. Centrifugal pumps work their best against a certain load. When starting a centrifugal pump, start it against a closed delivery valve which compares with the manufacturer's recommendation. On the delivery side, the rise lines lead from the cargo pumps to the main deck. The first is the cow crossover line. With this line, we can bleed off from any riser for supplying crude oil washing during discharging or supplying water during tank washing. The same line also supplies "drive" when using the ejector for stripping. The second crossover line leads to a higher inlet in the port slop tank (primary slop) and to the "high overboard" line. The high overboard line is the line where ballast water and washing water are discharged overboard via oily water detection and monitoring equipment. As the drawing illustrates, it is possible with any cargo pump to crossover to any of the risers. The pump room is also fitted with other equipment for handling cargo and ballast. The ballast pump is only used for segregated ballast. The segregated ballast system is totally isolated from the cargo systems. The ballast pump is connected to the FP tank and the WT3 S/P. The ballast system has its own sea chest. Still, there are some vessels, among them the MT SEAGULL, which have separated lines from the ballast pump to the main deck, which end in drop lines to the cargo tanks that are dedicated to departure and arrival ballast. These tanks can be ballasted without involving any part of the cargo line systems. The stripping pump operates its own system, which (via a stripping crossover) strips out the last amounts of cargo from the tanks, cargo pumps and lines, through a small diameter line and ashore. In addition to a stripping pump and an ejector, these vessels are equipped with a vacuum stripping system, which provides the cargo pumps with the ability to maintain suction when only a small quantity of cargo is left in the tank.

Deck lines

On crude tankers, the main line system changes name depending on where on the vessel it is located. From cargo tanks to the cargo pumps, the main lines are called "bottom lines." From the cargo pump delivery side, the name changes to risers. When they appear on the main deck, the names are deck lines. Very often the systems are numbered from one side of the ship to the other, for instance from port to starboard or vice versa. The deck lines are a lengthening of the risers from the pump room. Each deck line can be isolated to the pump room by the deck master valve. The deck lines end up at the manifold crossover lines. These manifolds are where the vessel is connected to the terminal by hoses, kickarms, etc. The manifold line is numbered with the same number as the main line it belongs to. This means manifold no. 1 is connected to drop line no. 1, which leads down to bottom line no. 1, which leads to cargo pump no. 1, which leads to riser no. 1, which leads to deck line no. 1, which leads to manifold no. 1. The same applies to system no. 2, no. 3, and no. 4. The vessel is also equipped with manifold crossovers, which makes it possible to operate between deck lines, drop lines and manifolds depending on which manifold(s) the vessel is connected to. By studying the ship's line system, including the myriad valves and crossovers, we can see there are all manner of possibilities for leading cargo or water through the ship's systems. The more we are familiar with the line system and its drawings, the better we can use the system's possibilities. On the main deck, we also find the small diameter line (or MARPOL line) which leads from the vessel's stripping pump to one of the vessel's manifolds. The small diameter line is connected on the outside of the manifold valve, which in turn is connected to the "presentation flange." The purpose of this line is to strip the last amount of cargo ashore from the tanks, pumps and lines. When using this line, it is important to keep the specific manifold valve closed, to avoid the cargo returning into the vessel's lines.

TYPES OF PUMPS AND EDUCTORS

A pump is used to move liquids from lower pressure to higher pressure. The main types of pumps, i.e., positive displacement and dynamic pressure pump types, found on oil and chemical tankers are listed in Tables 2.1 and 2.2.

Table 2.1 Positive displacement pump types

Reciprocating pump	Piston pump
Screw pump	Ram type pump
Gear pump	Vane pump

Table 2.2 Dynamic pressure pump types

Centrifugal pumps	Submersible pumps
Axial flow pumps	Centrifugal-axial (mixed) pump

Cargo and ballast pumps

As stated above, the function of any pump is to transfer liquid from one point to another and this involves the use of piping. Such transfers onboard tankers can be divided into two parts. The first part relates to the movement of liquid from the tank to the pump. This is a function of the pump and its installation design. These factors are beyond the control of the ship provided the design ratings of the pump are maintained. The second part relates to the onward movement of the liquid from the pump to its destination. This is an area where the efficient operation of the pumps is essential if optimum results are to be obtained. The flow of liquid to and from the pump must be matched exactly and this requires the flow on the suction side to be equal to or greater than the discharge rate of the pump. Where the flow to the pump suction falls below the pumping rate, cavitation will occur resulting in the potential for loss of suction and pump damage. Centrifugal pumps do not suck liquids. The only factors which cause liquid to flow to the pump are:

- Pressure acting on the surface of the liquid, and
- The height of the liquid level in the tank in relation to the pump suction. Since no centrifugal pump can generate a total vacuum at its suction inlet, only a proportion of the atmospheric pressure can be usefully employed.

Therefore, before a pump can operate satisfactorily, a certain pressure must exist at the pump suction. This is technically referred to as the required *net positive suction head.*

Centrifugal pumps

The centrifugal pump is a kinetic type pump which increases the flow rate of the liquid as passes through the pump. This centrifuges the liquid into the discharge line. If the flow into the pump suction is less than the delivery, the pump will cavitate causing gassing and loss of suction. Centrifugal pumps are resistible to the suspended solids found in oil and chemical cargoes. The primary disadvantage though of centrifugal pumps is their inability to evacuate air/gas from its casing. This means the pump casing must be filled with liquid before it is started. The pump must be stopped to do this. When the impeller starts to turn the liquid is driven to the periphery of the housing by centrifugal forces. This results in the development of positive pressure on the outside of the impeller and a negative pressure in the centre. The centrifugal pump has for many years been the most suitable pump where a high pumping capacity is the most important factor. Unlike positive displacement pumps, the size and cost of centrifugal-type pumps do not increase in proportion to the throughput. However, they do require either the provision of ancillary self-priming equipment for the removal of air in the system or a separate stripping system.

In a centrifugal pump, the motive force is provided by a rotating impeller which takes its suction at its centre and centrifuges the pumped liquid outwards to the casing discharge. The head generated is dependent on the diameter, blade angle, and speed of rotation of the impeller. The flow rate is affected by the pressure in the discharge system and, if not maintained properly, can fall to zero. Reverse flow through the pump can occur where a non-return valve is not fitted and operational on the discharge side of the pump. The correct and efficient use of centrifugal pumps requires the observance of certain basic operating principles. Whilst this book provides high-level guidance on the principles of operation of centrifugal pumps, it must be noted that manufacturers often incorporate special design features to meet operational requirements. Therefore, the information provided herein must be read in conjunction with the manufacturer's operating instructions and ship-specific onboard procedures. The basic characteristics of a centrifugal pump are:

- Throughput varies with speed,
- Head varies as speed squared, and
- Power required varies as speed is cubed.

These relationships are subject to appreciable variation caused by the system in which the pump operates.

Positive displacement pump

Unlike the centrifugal pump, the positive displacement pump used in dedicated stripping systems is capable of a low suction pressure and the ability to pick up suction without external priming. This type of pump includes steam reciprocating pumps and "screw" type pumps. Both types are now mainly used for stripping tanks or as specialised cargo pumps. The suction and discharge valves of a positive displacement pump must always be open before starting the pump and must remain open until the pump is stopped. These pumps must not be operated in excess of their design speed and particular care must be taken to avoid these pumps over-speeding when they lose suction. Pressure relief devices must be checked at regular intervals to ensure their correct operation.

Submerged pumps

Submerged pumps are common on chemical carriers. This type of pump is usually powered hydraulically or electrically and provides for a pump located in each tank. Manufacturer's instructions must be complied with for efficient operation of these pumps. Submersible pumps are purged, using inert gas (i.e., using the ship's inert gas (IG) or nitrogen) or air, as a means of checking for seal integrity condition and tightness. The pumps must be purged before and after every loading/discharging/tank cleaning operation and the appropriate record form(s) completed. If the purging records indicate a deviation from the manufacturer's recommended parameters, such as one or more pump cofferdams are blocked or excessive seal leakage is detected, the ship's management office must be notified immediately, and appropriate corrective action taken at the first opportunity.

Portable submersible (emergency) pump

Portable submersible pumps are often installed on chemical ships and other specialised liquid cargo carriers, for the purpose of discharging cargo in the event of a main cargo pump failure. The pumps are usually hydraulically driven and lowered directly into the tank through a tank cleaning hatch. All necessary safety precautions relevant to the specific cargo being handled must be observed and permission obtained from the local port authorities before operations can be commenced. It is good practice to shut down the hydraulic oil pressure system before connecting and disconnecting the hydraulic hoses of portable hydraulic-driven emergency pumps.

Use of eductors

Eductors may be used for ballast stripping purposes. To strip efficiently, an eductor used for tank cleaning operations should have a capacity of about twice the rate of liquid being introduced to the tanks. For safety and efficiency, eductors must always be operated at or near their design driving pressure as, in general, lower driving pressures will reduce the eductor's efficiency. Higher back pressures in the system than the eductor was designed for can also reduce suction capacity. The eductor drive liquid must always be flowing before the suction valve is opened. This will prevent the back flow of the driving liquid to the tank suction. When shutting down an eductor the suction valve is to remain open until the eductor is stopped to prevent the eductor drawing a vacuum on the suction line. If, during use, the eductor driving pressure falls below the required operating pressure, the eductor suction valve must be closed to prevent backflow of the driving liquid. The tank suction must not be used to prevent backflow as the suction pipework is not typically designed for such high operating pressures. High melting point cargoes such as phenol, palm fatty acid distillates, lauric and stearic acid have inherent properties which are liable to form lumps in the cargo. It is therefore recommended to turn the cargo pumps at regular intervals during the voyage, and prior to discharge, in order to avoid any potential blockages. All lines should then be blown back to the cargo tanks. The danger of frozen valves and pressure/vacuum in the tanks must be regularly monitored during loading, voyage and discharge. In the event any of the pumps are found to be frozen, the deployment of portable Framo pumps may be considered, pending proper risk assessment and authorisation.

Notes

1 SOLAS, MARPOL, STCW.
2 ISM.
3 For example, Lloyd's Register (LR); Bureau Veritas (BV); American Bureau of Shipping (ABS); DNVGL; RINA.
4 Before entering the pump room, always ensure it is completely gas-free. Never enter an enclosed space unless authorised and it is safe to do so.

Chapter 3

Physical and chemical properties of oil and chemical products

GENERAL

Basic physics (system International de Units [SI system of units])

In earlier days, many systems of units were followed to measure physical quantities. The British system of foot–pound–second or the FPS system, the Gaussian system of centimetre–gram–second or the CGS system and the metre–kilogram–second or the MKS system are just three of the most commonly known of the 17 different systems followed.

Mass and weight

The mass of an object is a fundamental property of the object; a numerical measure of its inertia; a fundamental measure of the amount of matter in the object. Definitions of mass often seem circular because it is such a fundamental quantity that it is hard to define in terms of something else. All mechanical quantities can be defined in terms of mass, length and time. The usual symbol for mass is m and its SI unit is the kilogram. While the mass is normally considered to be an unchanging property of an object, at speeds approaching the speed of light one must consider the increase in the relativistic mass.

The weight of an object is the force of gravity on the object and may be defined as the mass times the acceleration of gravity, $w = mg$. Since the weight is a force, its SI unit is newton. Density is mass/volume.

BASIC CHEMISTRY, CHEMICAL ELEMENTS AND GROUPS

Bonds and molecules

An atom stabilises by bonding with another atom in order to fill out its outer set of electrons in its shell. When two atoms of the same chemical

DOI: 10.1201/9781003505044-3

element bond together they form a diatonic molecule. When two atoms of different chemical elements bond, they form a chemical compound. Atoms are held together because there is an electrostatic attractive force between the two atoms. Energy is required for the chemical reaction to bond atoms. This energy becomes potential chemical energy that is stored in a molecule or chemical compound. For example, combining two atoms of hydrogen forms a hydrogen molecule, H_2. Combining a hydrogen molecule consisting of two atoms with one oxygen atom forms the compound we know as water (H_2O). Bonds are formed in two ways:

- Gain or lose an electron from the valence shell; called an ionic attraction, and/or
- Share one or more electrons in the valence shell; called a covalent bond.

Atoms bond together using a range of ionic and covalent bonds. There are three major chemical bonds:

- **Ionic bond.** Transfer electrons from one atom to another atom. An atom becomes unbalanced when it gains or loses an electron. The reaction that creates table salt from sodium and chlorine causes an ionic bond between these atoms,
- **Covalent bond.** Atoms share electrons in their valence shell. The shared electron orbits the nucleus of both atoms. A covalent bond is the strongest bond and the most commonly found in organisms, and
- **Hydrogen bond.** A hydrogen bond forms a weak, temporary bond that serves as a bridge between either different molecules or portions of the same molecule. For example, two water molecules are physically combined using a hydrogen bond.

Compounds

Atoms are seldom solitary in nature. They tend to combine with other atoms to create compounds, also known as molecules. Common examples of molecules include water, carbon dioxide and ammonia. Molecules are typically represented by molecular formulas which list the number and type of atoms found in the molecule. Listed in Table 3.1 are the molecular formulas of some of the most common water treatment chemicals.

Periodic table of groups of elements

One reason the periodic table of elements is so useful is because it is a means of arranging elements according to their similar properties. There are multiple ways of grouping the elements, but they are commonly divided into metals, semimetals and nonmetals. You will find more specific groups, like

Table 3.1 Common chemical names and formulas

Common name	*Chemical name*	*Formula*
	Aluminium hydroxide	$Al(OH)_3$
Filter alum	Aluminium sulphate	$Al_2(SO_4)_3 * 14H_2O$
Ammonia gas, ammonia	Ammonia	NH_3
Ammonia	Ammonia hydroxide	NH_4OH
	Calcium bicarbonate	$Ca(HCO_3)_2$
Limestone	Calcium carbonate	$CaCO_3$
Hydrated lime	Calcium hydroxide	$Ca(OH)_2$
HTH	Calcium hypochlorite	$Ca(OCl)$
Quick lime	Calcium oxide	CaO
Chloride of lime	Calcium oxychloride	$CaOCl_2$
Gypsum	Calcium sulphate	$CaSO_4$
Dry ice	Carbon dioxide	CO_2
	Carbonic acid	H_2CO_3
Liquid chlorine	Chlorine	Cl_2
	Chlorine dioxide	ClO_2
Blue vitriol	Cupric sulphate	$CuSO_4 * 5H_2O$
	Dichloramine	$NHCl_2$
Ferrichlor, chlorine of iron	Ferric chloride	$FeCl_3 * 6H_2O$
Muriatic acid	Hydrochloric acid	HCl
	Hypochlorous acid	$HOCl$
	Methane	CH_4
	Monochloramine	NH_2Cl
	Nitrogen trichloride	NCl_3
Soda	Sodium bicarbonate	$NaHCO_3$
Soda ash	Sodium carbonate	Na_2CO_3
Salt	Sodium chloride	$NaCl$
Lye, caustic soda	Sodium hydroxide	$NaOH$
	Sodium phosphate	$Na_3PO_4 * 12H_2O$
Oil of vitriol, vitriol	Sulphuric acid	H_2SO_4
Water	Water	H_2O

transition metals, rare earths, alkali metals, alkaline earths, halogens and noble gases. Click on an element to read about the chemical and physical properties of the group to which that element belongs.

Physical properties of various noxious liquid chemicals carried at sea specific gravity (density)

Tanks on a Chemical Tanker are normally designed to load cargoes of a higher specific gravity than an oil tanker. Very often the design strength differs between groups of tanks on the same ship. The information with regards to tank strengthening is normally found on the Certificate of Class and Fitness, and the Master must be familiar with this layout and the restrictions that may be imposed on loading high gravity cargoes. Especially important is the risk of slack loading a tank as this can cause excessive sloshing in the tank that may cause damage to the tank structure and/or its equipment. Equally important is the danger of exceeding the tank's design weight capacity.

Flash point

The flash point of a liquid is the lowest temperature at which the liquid will give off sufficient vapour to form a flammable gas mixture with air, near the surface of the liquid.

Auto-ignition temperature

The auto-ignition temperature of a solid, liquid or gas is the lowest temperature at which it requires to be raised to support self-combustion.

Flammable/explosive limits

The flammable limits (explosive limits) are the minimum and maximum concentrations of flammable gas or vapour in air between which ignition can occur. The minimum vapour concentration is known as:

- The Lower Flammable Limit (LFL), or
- The Lower Explosive Limit (LEL).

The maximum vapour concentration is known as:

- The Upper Flammable Limit (UFL), or
- The Upper Explosive Limit (UEL).

Vapour pressure/boiling point

The vapour of every liquid exerts a certain vapour pressure at any given temperature called the vapour pressure. The liquid will boil when the vapour pressure equals the external atmospheric pressure. In a closed ship tank, however, the liquid will boil when the vapour pressure equals the atmosphere pressure plus the pressure setting of the P/V valve. The tanks and vent systems are designed to withstand this pressure, plus the hydrostatic pressure of the cargo.

True vapour pressure (TVP)

The true vapour pressure of a liquid is the absolute pressure exerted by the gas produced by evaporation from a liquid when gas and liquid are in equilibrium at the prevailing temperature. Boiling Point The temperature at which the vapour pressure of a liquid equals that of the atmosphere above its surface; this temperature varies with pressure.

Freezing point/melting point

Most liquids have a defined freezing point, sometimes described as the melting point. Some products, like lube oil additives, vegetable and animal oils and polyoils, do not have a defined freezing point, but rather a freezing (melting) range or none at all. The product's viscosity is instead used as a measurement for the product's liquidity or handling characteristics. Products with a freezing point higher than the outside temperature in which the ship is trading will need to be heated in order to remain liquid. Ship's structure and equipment normally have limitations on high heat. Exceeding this limitation could damage the tanks or their structure. High heat will also reduce steel strength, and the risk of cracking will increase. Caution should be exercised when carrying high-heat products as non-insulated lines and vents may freeze and clog the systems. Not insulated cargo lines used for high-heat products pose a safety hazard as they may cause severe burns if touched. Adjacent tank temperature limitations must be actively monitored. MARPOL Annex II requirements for solidifying substances discharge temperature to be complied with (in consultation with shippers). Prewash may be required if discharge temperature as per Annex II cannot be complied with. The cargo tank vapours pressure to be monitored carefully in freezing weather conditions to monitor blockage of PV vent lines.

Solidifying/non-solidifying properties

Solidifying substance means a noxious liquid substance which:

1. In the case of a substance with a melting point of less than 15°C (59°F) which is at a temperature of less than 5°C (41°F) above its melting point at the time of unloading; or
2. In the case of a substance with a melting point of equal to or greater than 15°C (59°F) which is at a temperature of less than 10°C (50°F) above its melting point at the time of unloading.

Non-solidifying substance is a noxious liquid substance, which is not a solidifying substance.

Pour point

The pour point of a liquid is the lowest temperature at which the liquid will flow. It should be noted that cargo with thixotropic properties (the properties of showing a temporary reduction in viscosity when shaken or stirred) can be pumped at temperatures well below its pour point but at very restricted rates.

Viscosity

Viscosity is a measure of a liquid's ability to flow and is usually determined by measuring the time required for a fixed volume to flow under gravity through a thin tube at a fixed temperature. As the temperature of the liquid increases its viscosity decreases and therefore it flows more readily. It can also be described as a measure of the internal friction of a liquid. The distinction between viscosity and pour point should be made clear. Oil ceases to flow below its pour point temperature when the wax content solidifies. A viscosity measurement of a liquid depends upon the internal resistance of the liquid to flow. For a simple liquid, this internal resistance varies with the temperature in a predictable and regular way. However, when oil approaches its pour point the rate at which viscosity increases as temperature falls accelerates until sufficient wax has precipitated to solidify the product. Viscosity is important as regards the pumpability of a product. Centrifugal and deep well pumps are acceptable for the majority of cargoes but high viscosity products such as bitumen or molasses are more suited for pumping with positive displacement pumps. High-Viscosity Substance is a noxious liquid substance in Category X or Y with a viscosity equal to or greater than 50 mPa at the unloading temperature. Low-Viscosity Substance is a noxious liquid substance, which is not a High-Viscosity Substance. MARPOL Annex II requirements for high-viscosity substances are to be complied with.

Electrostatic charging

Certain cargoes are known as static accumulators and become electrostatically charged when handled. They can accumulate enough charge to release a spark that could ignite a flammable tank atmosphere.

Cubic expansion

All liquids will expand as temperature rises or contract when temperature decreases. Sufficient space must be allocated in the tank to facilitate any cubic expansion or contraction expected during the voyage. Vent line systems must be checked for operation at regular intervals, as a malfunction could cause structural damage because of changes in the liquid's volume. For calculating maximum intake, the density at 35°C (95°F) is used for non-heating cargoes and the density at maximum discharge temperature for heated cargo is used. The volume at these temperatures should not exceed 98% of the cargo tank's maximum volume. Allowances should also be made for load density and the International Bulk Chemical Code (IBC Code) requirements.

Vapour density

Vapour density is expressed relative to the density of air. Many chemical cargo vapours are heavier than air, caution must be exercised during loading and any other cargo operation, as vapour concentrations may accumulate and be trapped in certain deck areas. (If cargo tanks are incorrectly cleaned vapour may remain in the bottom of the tank).

Solubility

Solubility is expressed in many different ways; yes, no, slight, as a percentage or totally and in this connection only with water. Most non-soluble chemicals are lighter than water and will float on top; others like the chlorinated solvents are heavier and will sink to the bottom. This latter condition may cause a safety risk in drip trays and even in cargo tanks where they may be trapped under water in pump wells and pose a danger even if the tank atmosphere is tested safe for entry.

Colour

Colour is the comparison between a sample of product and standard colours measured under tightly controlled conditions. The colour of clean products is one of the more common causes of cargo rejection or downgrading. This is caused by loading a light-coloured product without adequate preparation into a tank that last carried a darker product. Most of the lube oils and white-water–white products show quite readily the traces of prior darker lube oils or residual products, and because of this trait, it is most important that the tank cleaning instructions are closely followed, and proper line cleaning is carried out.

Chapter 4

Cargo operations

GENERAL

Cargo operations onboard oil and chemical tankers require extra and special care as the nature of hazards are different and are in addition to the hazards found onboard other types of vessels. The main hazards associated with oil and chemical tanker cargo operations include fire and explosions, toxic spillages, corrosiveness and reactivity of the cargo with air, coatings, metals, water, other cargo and itself. Safety notices must be placed at conspicuous places with warning signs as applicable. **SAFETY FIRST** must always be kept in mind at all times. Before berthing the vessel, complete information must be obtained on the rules and regulations applicable at the terminal of arrival (whether loading or discharging) and any extra procedures to be followed pursuant to the ship's standard operating procedures (SOP). Before any cargo is loaded onboard, a complete specification of the cargo and its appropriate Material Safety Data Sheets (MSDS) must be obtained. The cargo loading plan must also be prepared in advance of arrival. It is also worth reviewing the ship-shore checklist in good time before they must be completed. Only competent personnel are to be authorised to undertake loading and discharging operations. In all cases, the tanks must be inerted as required and oxygen levels maintained. Cargo must be loaded in accordance with the loading plan to ensure proper segregation, coating compatibility etc. Safe loading rates must be followed and controlled. On completion of loading, and during voyage, the correct cargo temperature for each cargo must be maintained. Discharge operations require similar precautions with special care to avoid spillages due to improper coordination with the shore terminal, for example, accidental closing of shoreline valves.

DOI: 10.1201/9781003505044-4

OFFICER AND CREW RESPONSIBILITIES

Responsibility of the Master

The Master has the whole responsibility for cargo handling operations as outlined throughout this book. As such, the Master must ensure that the crew under their command are familiar with all aspects of their respective duties. In supervising cargo operations, the Master must be conversant with the contents of all relevant laws and regulations, charter parties, sailing/shipping instructions etc. Before approving the cargo handling plan as prepared by the Chief Officer, the Master must check and evaluate the appropriateness of the plan in accordance with the statutory laws and regulations and the regulations of the terminal(s) concerned. During cargo handling operations, the Master must assign officers and crew to maintain a proper watch. The Master must hold a meeting before arrival at port and disseminate the cargo handling plan and safety measures among the crew members so that each member of the ship's crew is thoroughly familiar with them. Finally, the Master must obtain all the information in the form of cargo data sheets, etc., from the shippers; this data must then be communicated to all ship's staff.

Responsibility of the Chief Officer

The Chief Officer is personally responsible to the Master for safe and efficient cargo handling operations. It is the Chief Officer's responsibility to command and supervise the operations under the direction of the Master and on the basis of the approved cargo handling plan. The Chief Officer must post Cargo Data Sheets in conspicuous spaces for all on board to refer to. The Chief Officer must further familiarise themself with the properties of cargo and prevent the occurrence of accidents and should be thoroughly familiar with the antidotes for the cargo carried in case of personal contact. The Chief Officer must post the cargo operation plan in the cargo control room (CCR) and give necessary orders to other deck crew members in writing, or verbally, thereby accomplishing the operation safely and efficiently. Before commencing cargo operations, the Chief Officer must direct other deck officers and crew to check the cargo handling equipment for proper operation and maintenance, and effect readjustment, if necessary. In the event the ship is capable of carrying special cargoes as per Chapter 15 of the IBC Code, then the Master and Chief Officer must refer to this chapter prior to approving the cargo plan. All necessary precautions must be complied with. Note that Chapter 15 cargoes are dangerous and require special precautions. In cases where the Master is not satisfied with the information provided by the shipper with respect to the properties of the cargo to be loaded, the precautions required for transportation, suitability for the type of coating required, or indeed any other special precautions which may be required, then the Master may exercise their right to request that related information from the shippers/charterers. In the event said information is not provided, the Master has the right to refuse the loading of such chemicals onboard the vessel. This must be reported to the vessel owner and the shipper immediately. Furthermore, the Chief Officer must establish a watch arrangement during all cargo work. This must be clearly communicated to all crew members. A copy of the watch arrangement should also be posted in the CCR.

Responsibility of the duty officer

The duty officer must thoroughly understand the cargo pumps, pipelines and other cargo handling equipment and instruments, deck auxiliary machinery, firefighting and lifesaving equipment as well as cargo handling operations on their vessel. The duty officer or their substitute should enter the pump room at least once every hour during cargo handling operations to check the operational condition of each pump and to measure the pressure and temperature of each pump at the same time. The results of the duty officers' observations should be recorded accordingly. For the pump room entries, the pump room entry permit must be complied with. The duty officer must assign crew members, upon commencing cargo or ballast water handling operations, to monitor the sea surface around the vessel in order to prevent oil leaks and/or oil spills. In addition, the duty officer must check for the presence of leaks at regular intervals during cargo handling operations. The duty officer must not fail to take due care by constantly directing and supervising deck ratings on the watch so that cargo handling operations may progress safely and efficiently. Ratings should also be posted them at specified stations in order to act required in response to any emergency situation. Should any problem beyond the duty officer's judgment occur, the Chief Officer must be informed immediately who will advise on the appropriate actions to be taken. The duty officer must make efforts to maintain proper communication with the duty engineer (or the person on unmanned duty) so that the cargo operation can be accomplished safely and efficiently.

CATEGORISATION OF CARGOES

Chemical cargo refers to the cargo described in the "IMO Certificate of Fitness for the Carriage of Dangerous Chemical in Bulk" and is categorised as follows.

Corrosive liquids

Acids, anhydrides and alkalis are among the most commonly carried corrosive substances. They can rapidly destroy human tissue and cause irreparable damage. They can also corrode normal ship construction materials and create a safety hazard for a ship. Acids in particular react with most metals, evolving hydrogen gas which is highly flammable. The International Maritime Organisation (IMO) Codes address this, and care should be taken to ensure that unsuitable materials are not included in the cargo system. Personnel likely to be exposed to these products should wear suitable personal protective equipment.

Acids

In chemical terms, an acid is a substance containing hydrogen which, when dissolved in water, becomes dissociated and generates hydrogen ions. In high concentrations, many inorganic (or mineral) acids passivate mild steel and corrode it. But if the acid is diluted with water, rapid corrosion will occur. The most corrosive concentrated acid cargoes include nitric acid, sulphuric acid, chlorosulphonic acid and chloropropionic acid. Formic acid and acetic acid are also highly corrosive in concentrations above 90%. Some acids are called fuming acids because of their characteristic appearance as they give off corrosive acid vapours, during which large quantities of acidic fumes are generated. Acids can also have other dangers. Nitric acid is a powerful oxidising agent. It can cause a fire in contact with combustible materials; therefore, materials such as sawdust and cloth should never be used to collect spilt nitric acid or other oxidising agents. Sulphuric acid and chlorosulphonic acid react violently with water; the reaction gives off large amounts of heat which causes the water to boil. Some acids are toxic as well as corrosive and can cause damage to the body as well as the acid burn at the point of contact.

Alkalis

Alkaline substances are those which contain the oxidrile group OH^-. When dissolved in water, basic substances get dissociated and generate OH^- ions. Alkaline solutions contain such OH^- ions in higher concentration than pure water. For example, Sodium Hydroxide (caustic soda, NaOH) is an alkaline substance; dissolved in water it gets disassociated into Na^+ and OH^- ions. Common inorganic alkalis such as potassium hydroxide and sodium hydroxide (caustic soda) are corrosive to aluminium, zinc, galvanised steel and mercury, so those materials must not be used for cargo containment systems when carrying such chemicals. Most Alkalis have corrosiveness as either the primary risk or the secondary risk after flammability. Acids and Alkalis react together to form salts and water, often with a violent emission of heat.

Toxic cargoes

A toxic substance is one which is liable to cause either harm to human health, serious injury or death. Toxic means the same as poisonous. Toxicity is an intrinsic property of a chemical, which humans cannot modify, and its effect is a function of exposure. In some cases, the correct response to its effects after exposure can diminish its consequences. There are three common ways that cargo can be toxic: swallowed (oral toxicity), absorbed through the skin, eyes and mucous membranes (dermal toxicity) or inhalation as a vapour or mist (inhalation toxicity). A chemical may be toxic by more than one of these routes: for example, toxic vapours and mists affect people most via the respiratory system but they can also be absorbed through the skin. The smaller the quantity (or dose) of the substance that is required to harm health, the more toxic a substance is. In some cases, the toxic effect of a chemical can be countered by administering antidotes, but in most cases, the hazard must be avoided by correct use of protective clothing, breathing apparatus and ventilation procedures. If there is no exposure to the chemical, or if exposure is reduced to safe levels, there can be no toxic effect. In tanker operations, contact with a liquid or inhalation of a vapour are the forms of exposure. In general, proper procedures and proper use of personal protective equipment will prevent exposure and thus the effects of toxicity.

Toxic effects

Toxicity can be acute, sub-acute and chronic. A substance has acute toxicity if a single exposure is sufficient to cause harm almost immediately. Substances commonly called poisons have extreme acute toxicity. A substance with sub-acute toxicity displays its effects after a person has had repeated exposures to doses too small to cause an acute effect. Examples are allergic sensitisers, which induce reactions to other substances. A substance has chronic toxicity if its effects appear after a period of continuous exposure to doses too low to cause any acute effect. Examples are carcinogens (cancer-inducing), teratogens and mutagens (which affect reproduction).

Threshold limit value (TLV)

A threshold limit value (TLV) for a given substance is the maximum concentration of its vapour in the air to which it is believed that personnel may be exposed under certain circumstances without suffering adverse effects. Various governmental bodies publish TLVs. These should not be regarded as the absolute dividing line between safe and hazardous conditions. It is a good operating practice to keep all vapour concentrations to a minimum and a safe margin below the TLV. The best-known list of TLVs is issued by the American Council of Governmental Industrial Hygienists (ACGIH). The values are updated annually in the light of new knowledge, so it is important to refer to the latest edition. The ACGIH defines three categories of TLVs:

- TLV – TWA (Time Weighted Average). The concentration of vapour in the air may be experienced for an eight-hour day or 40-hour week throughout a person's working life. This is the most commonly quoted TLV.

- TLV – STEL (Short Term Exposure Limit). The maximum concentration of vapour in air is allowable for a period of up to 15 minutes, provided that there are not more than four exposures per day and at least one hour between each. It is always greater than the TWA. It is not given for all vapours.
- TLV – C (Ceiling). An absolute maximum which should never be exceeded. It is given only for fast-acting substances. This is the highest of the three values for a given substance.

Precautionary principles

Containment is the first objective when any toxic substances are handled, by making sure that they stay inside the cargo system. Engineering and ship design features will provide a secure storage space. If there is no exposure there is no toxicity danger, however hazardous the chemical can be. Leakage of liquid or release of vapour must be prevented by keeping the cargo system closed unless it is absolutely unavoidable to open it. However, some operations inevitably involve opening the system; for example, disconnecting a hose from the ship's manifold after the transfer of cargo. Although this is a routine operation, it should be regarded as – comparable to opening up a cargo line elsewhere on deck, and operators must wear the necessary personal protective equipment.

Toxic vapour detection and personal protective equipment

Most chemical vapours are heavier than air and tend to flow along the deck and accumulate in low spots, for example below pumproom floor plates. Therefore, atmosphere samples should always be taken in such low points where concentrations are likely to be highest. It is important that a full chemical suit is worn by personnel when:

1. Inspecting pipelines and machinery for leaks,
2. Dealing with accidental leaks and spillage,
3. Connecting and disconnecting hoses and loading arms,
4. Taking ullages and samples from tanks (where restricted gauging is permitted),
5. Entering enclosed spaces such as pumprooms, cofferdams and tanks unless certified gas-free, and
6. Opening up pumps and equipment (unless certified gas-free).

IBC CODE REQUIREMENTS

The IBC Code specifies ways to limit the exposure of personnel to toxic vapours while cargo is being handled or during carriage at sea. First, it minimises toxic vapour emissions by controlling how cargo vapours are to be vented or returned to shore and how tank contents are to be gauged. All toxic cargoes require closed or restricted tank gauges to prevent crews from being exposed to unsafe concentrations of toxic vapours. Second, it specifies ventilation of working spaces such as pumprooms, requires the ship to carry equipment to detect vapours, requires the provision of personal equipment and, to ensure that toxic vapours are diluted to safe concentrations before they can reach accommodation areas, requires that tank vent system outlets are located at a safe distance (i.e., the safe distances specified depend on the severity of the toxic hazard). Third, it reduces the likelihood of accidental overflow spills by specifying that all acutely toxic products and all allergic sensitisers are to be carried in tanks equipped with a visual and audible high-level alarm (HLA). Tanks certified for the most severely acute toxic products must have a further overflow control system.

Finally, it specifies that cargo piping, including pumps, and venting systems of tanks carrying toxic cargoes are to be separated from those containing other products, to prevent any leakage causing toxic contamination of non-toxic products and subsequent exposure of personnel unaware of the contamination. This is achieved on many chemical tankers by having separate pumps, pipelines and vents so that segregation is achieved by design, and on ships with common pipeline systems by the engineering principle of two physical stops, such as spectacle plates or a removable spool piece and blank flanges. Valve gland packing is the source of many small leaks. The correct packing material for the chemical being carried should always be used, and the glands correctly tightened. The IBC Code prohibits the stowage of most toxic products adjacent to oil fuel tanks. The combustion of many otherwise non-toxic chemicals may produce toxic substances such as carbon dioxide and carbon monoxide, fumes of hydrochloric acid, hydrogen cyanide and nitrogen oxides. These may be present at some distance from the fire and may have no warning odour. Self-contained breathing apparatus should be used when dealing with chemical fires. The main danger from fume inhalation is asphyxia. Personnel affected by fumes should be removed rapidly to a fresh atmosphere, given oxygen and then treated appropriately as shown in the Medical First Aid Guide (MFAG).

MEDICAL FIRST AID

The two fundamental guides for medical first aid onboard ships, which provide advice on responding to exposure to toxic cargoes, are the International Medical Guide for Ships (IMGS) and the MFAG for Use in Accidents Involving Dangerous Goods (MFAG). Both are published jointly by the IMO, ILO (the International Labour Organisation) and WHO (the World Health Organisation). The IMGS gives guidance on common illnesses and is not solely concerned with chemical accidents. The MFAG is supplementary to the IMGS and contains advice for recognising and treating chemical poisoning, within the limits of the facilities available on board. The general rule is that if, during the handling of chemicals, any person shows symptoms that suggest poisoning, they should be treated in accordance with the MFAG and seen by a doctor as soon as practicable. Medical advice should be sought by radio, while still at sea. Assistance may also be available from another ship with a doctor on board.

Emergency treatment according to the route of exposure

Note that over and above the routine blood tests, as required under the crewing procedures, any crew member exposed to toxic chemicals must have their blood tested immediately after and then again as per a recognised physician's advice. MFAG gives directions on how to recognise the general symptoms of poisoning. Note that they may not appear for some time after exposure to the chemical. Symptoms to be alerted to are unexpected headaches, nausea and vomiting, drowsiness, changes in mental behaviour, unconsciousness, convulsions or pain. If the patient has a rapid but weak pulse, a greyish-blue colour of the skin, severe breathing difficulty or remains unconscious for a prolonged period, severe poisoning must be suspected.

First aid and further care

A first aider is not just a person with goodwill, but a person with training. MFAG outlines immediate first aid, i.e., treatment for minor casualties, or to enable a victim to be moved so that further treatment can be administered. The key priorities to remember when reacting to a casualty are:

- Send for help and inform the Master,
- Avoid becoming the next casualty: if the response is too big for you alone, wait for backup,
- Remove the casualty from the danger or vice versa, and
- Use a breathing apparatus if there is any suspicion of toxic gases or vapours in the area.

The signs and symptoms of mild poisoning usually resolve after a few hours in the majority of incidents, particularly if the degree of exposure is small. However, if a greater amount is taken in, or the period of exposure is prolonged or the chemical is very toxic, symptoms may persist for much longer, even for some days. The patient's condition may continue to deteriorate even when clear of the source of the vapour, and systemic effects may appear. Finally, the warning is given that death may occur despite treatment. General advice can be found in MFAG, on the emergency treatment to be administered according to the way the chemical has entered the body, for example by skin or eye contact, ingestion or inhalation. If the chemical has affected both the skin and the eyes, the latter should get priority for attention. If the chemical has been ingested, the patient should not be made to vomit because the vomit may enter the respiratory system and add to the exposure problem.

MFAG tables

Appropriate reactions after exposure to the toxic cargoes listed in the IMO Codes are given by tables in MFAG. There are 12 group tables, but five chemicals require their own single substance tables because they present particular combinations of toxic hazards (carbon disulphide, allyl alcohol, benzene, acrylamide and tricresyl phosphate).

Emergency schedules

The Emergency Schedules are an appendix to the International Maritime Dangerous Goods (IMDG) Code and are written to provide the vessel Master with advice on the immediate action to be taken in case of accidents such as spillage or leakage of toxic substances. In brief, if it is safe to do so, spillage should be collected for subsequent disposal, but if there is any doubt, the spillage should be washed overboard with plenty of water, because the safety of the crew takes priority over pollution avoidance.

REACTIVE/SENSITIVE CARGOES

Reactive/sensitive cargoes refer to liquids which cause the following phenomena by self-reaction and reaction with water, air and other substances etc.:

1. Generation of heat,
2. Generation of vapour/gas,
3. Rise of tank pressure,
4. Deterioration of cargo liquids,
5. Fire or explosion,
6. Health hazards,
7. Polymerisation (self-reaction), and
8. Cause pressure to rise in the tank.

A chemical may react in a number of ways; with itself, with water, with air, with other chemicals or with other materials.

Self-reaction

The most common form of self-reaction is polymerisation. Polymerisation results in the conversion of gases or liquids into viscous liquids or solids. It may be a slow, natural process which only degrades the product without posing any safety hazards to the ship or the crew, or it may be a rapid, exothermic reaction evolving large amounts of heat and gases. The heat produced by the process can accelerate it. Such a reaction is called a run-off polymerisation that poses a serious danger to both the ship and its personnel. Products that are susceptible to polymerisation are normally transported with added inhibitors to prevent the onset of the reaction. An inhibited cargo certificate should be provided to the ship before a cargo is carried. The action to be taken in case of a polymerisation situation occurring while the cargo is on board should be covered by the ship's emergency contingency plan.

Reaction with water

Certain cargoes react with water in a way that could pose a danger to both the ship and its personnel. Toxic gases may be evolved. The most noticeable examples are the isocyanates; such cargoes are carried under dry and inert conditions. Other cargoes react with water in a slow way that poses no safety hazard, but the reaction may produce small amounts of chemicals that can damage equipment or tank materials or can cause oxygen depletion. Certain chemical cargoes, mostly ethers and aldehydes, may react with oxygen in the air or in the chemical to form unstable oxygen compounds (peroxides) which, if allowed to build up, could cause an explosion. Such cargoes can be either inhibited by an anti-oxidant or carried under inert conditions.

Reaction with other cargoes (incompatible chemicals)

Some cargoes react dangerously with one another. Such cargoes should be stowed away from each other (not in adjacent tanks) and prevented from mixing by using separate loading, discharging and venting systems. When planning the cargo stowage, the Master must use a recognised compatibility guide to ensure that cargoes stowed adjacent to each other are compatible. As per the IMO code requirements (refer to Appendix C.5.2 of the IBC Code), incompatible chemicals must be kept strictly separated from each other throughout the cargo containment and handling system, in order to avoid accidental mixing. Separation should be achieved by having two barriers between the containment systems of the incompatible chemicals. The tank should be separated by a cofferdam, by an empty tank, a void space, a tank containing mutually compatible cargo or a piping tunnel. The piping or venting system for incompatible cargoes should be separated by removing a valve or spool piece and blanking off the exposed pipe ends or installing two spectacle flanges with a bleeder or equivalent means to detect leakage in the pipe between the spectacle flanges. Cleaning a tank and related cargo handling systems should be performed thoroughly if consecutive cargoes are incompatible. It cannot be stressed too strongly that parcels of chemicals should not be accepted for shipment or loaded onto the ship unless positive assurance is available that the various chemicals are compatible with the segregation capability of the vessel. The ultimate responsibility for the safety of the ship lies with the Master, who should ensure that the cargo distribution proposed for a voyage provides proper segregation or compatibility, using the data sheets of all chemicals to be loaded. If the data sheets fail to provide the necessary information, the Master should defer loading the cargo until consultation with the owner or other authority has produced satisfactory assurance that the proposed segregation plan for the cargoes to be loaded is safe.

USCG compatibility chart

Several authoritative bodies have divided chemical cargoes into groups, defined criteria for incompatibility between groups, and have published a list of incompatible cargoes. The most familiar is published by the US Coast Guard. According to this source, a mixture of two chemicals is considered hazardous (and the chemicals in question declared incompatible) when, under specified test conditions, the temperature rise of the mixtures exceeds 25°C (77°F) or a gas is evolved. Whether cargoes within a pair of groups are incompatible is indicated in a table, known as the compatibility chart. It is important to note that, while the table gives general indications, the footnotes and data sheets for two particular cargoes should always be consulted because there are a number of exceptions to the compatibility chart.

Reaction with other materials

The materials used in the construction of the cargo systems must be compatible with the cargo to be carried, and care must be taken to ensure that no incompatible materials are used or introduced during maintenance (e.g., by the material used for replacing gaskets). Some materials may trigger a self-reaction within the product. In other cases, reaction with certain alloys will be non-hazardous to the ship or crew but can impair the commercial quality of the cargo or render it unusable.

Heat adjacency

The maximum temperature of adjacent cargo permitted for each cargo to be loaded shall be obtained from shippers when handling heated cargo. In addition, care shall be taken to avoid indirect heating of adjacent cargoes and bulkheads during hot water washing of adjacent tanks.

Precautions for handling reactive liquids

Materials of cargo tank and cargo handling equipment must be in conformity with the metal materials described in the associated "Material Safety Data Sheets." The maximum temperature of liquids should be controlled by checking their temperature regularly. Self-reactive liquids should not be loaded adjacent to tanks containing liquids which need to be heated. Substances which promote self-reaction of cargo liquids should be completely removed from tanks to be loaded with reactive liquids. When handling liquids with added anti-polymerising agents, the manufacturer's guide for using anti-polymerising agents should be consulted and its usage and health hazards well understood. Liquids having dangerous reactivity properties with each other must never be loaded adjacently. They should be segregated from each other by means of a cofferdam or other void space. Tanks loading liquids which have dangerous reactivity with each other should be equipped with an independent vent system. Chemical products should never be loaded unless the Master clearly assures themself that they are able to be loaded within the ship's standard segregation system, causing no dangerous reaction amongst them.

Animal, fish and vegetable oils and fats

These are oils and fats extracted from animals, fishes and vegetables. They do not have any particular danger but some of them have the capability of absorbing oxygen and resulting in oxidisation. Due to this reason, the cargo tanks which have been discharged of these cargoes will be deficient in oxygen due to the clinging of cargo in the tanks.

Precautions for entering tanks after discharge of animal, fish and vegetable oils and fats

No entry into tanks may be allowed without the Chief Officer's express permission. Before entering tanks, the safety of entry into the tank should be confirmed by checking the oxygen content in the tank. Checking of oxygen content in tanks must be done at appropriate intervals during the period of work inside the tank. Personnel must always be positioned on deck near the tank hatches during all work inside the tank. Proper communication must be maintained between the crew working inside the tank and those positioned on deck. An enclosed space entry permit must be completed and fully complied with.

High melting point cargoes/high viscosity cargoes/solidifying cargoes

High melting point cargoes are heated to reduce "unpumpables" and to reduce the load on the ship's centrifugal pumps. They should be washed at a temperature above MP, with any cold interface removed. Chemicals have different thermal properties. Some will heat up very quickly, whereas others will require substantial heating before any temperature gain is noticed. Similarly, those cargoes that heat quickly tend to cool down quickly, whereas the cargoes which heat slowly have far better heat retention properties. Colour also matters; white is a poor absorber and a poor radiator, while black is good at absorbing and radiating heat. These individual cargo properties should be recognised when heating cargoes on the loaded voyage. The length of the voyage must also be considered when deciding on heating schedules to avoid waste of fuel. Unless instructions are given to the contrary, any "solidifying substance" should be maintained and discharged at a temperature of at least 10°C (50°F) above its melting point, to avoid prewashing. Stowage of cargoes with MP > 55°C (151°F) must **NOT** be adjacent to a cold interface lest the whole bulkhead forms a solidified thick film. Should the cargo require it, heating coils can be opened to the tank, when the liquid is at a level of approximately 1 metre (3.2 feet). This can offset the cumulative list at berth such as for cargoes like tallow. If a cargo that requires heating is deemed dangerous and/or toxic and could result in fumes and vapours leaking into the engine room, then an additional dangerous/toxic return tank is required in accordance with the BCH/IBC Code.

When loading solidifying cargoes, due regard must be made to ensure that lines do not become frozen. In any case, should stress and list/ trim conditions allow it, the tanks furthest away from the manifold should be completed first, using a sequence of working towards the shortest pipe length. In manifolds with connections to both port and starboard, the opposite side to loading gets frozen. This must be unblocked at sea before discharge port. Many company policies require that no tank be loaded to a volume greater than 98% at any time. After loading high MP cargoes, leave the pipelines open till the contents are drained properly. If proper written heating instructions are not received, the Master lodges a Note of Protest and informs the chemical operator immediately. If the tank is not blanketed with inert gas (such as nitrogen) the sedimentation or build-up at the tank bottom must be ascertained with a sounding rod. The charterer's heating instructions must be obeyed, and in any case, the heating increase must not exceed 5°C (41°F) in any 24-hour period. During cold weather, the functioning of P/V relief valves should be checked regularly. It is possible that humid air vented from a cargo tank may condense and freeze thus inhibiting ventilation. This is also possible for cargoes with a high melting point, such as phenol, where cargo vapours can solidify in the vent line. For sediment/high MP cargoes, ensure that the deep well impeller is not frozen by running the pump before the arrival port with the delivery valve shut.

When discharging a homogeneous heated cargo, the wing tanks should, as far as possible, be emptied before the centre tanks. During the discharge, several tanks may be emptied simultaneously, however, when the cargo sounding falls to around 1 metre (3.2 feet) the discharge of the tank must be ceased temporarily. Final emptying, stripping and sweeping of the tanks must then happen one tank at a time. The temperature in the tanks must be monitored and the heating must be adjusted in the tanks accordingly. Never leave high MP cargoes inside the pipeline even for short periods. When vessels are carrying solidifying products it is important that the products are at the maximum discharge temperature (or slightly above) at least one week prior to arrival at the discharge berth. Also, during this period, once the product is up to the required temperature, soundings must be made in each tank using a rod and line to determine if there are any solid products on the tank bottom. In addition, any ballast in double-bottom tanks beneath the cargo tanks **MUST** be lowered so that there is no direct contact between the ballast and the cargo in adjacent tanks.

Be aware of the slope of the pump delivery deck pipes towards the manifold. Some shipyards deliberately tilt the aft of manifold pipelines to avoid fitting super strip lines. When loading high MP cargoes in winter from barges, any delay between two barges can freeze cargo inside the pipelines. In the event the cargo falls below the steam coils, the surveyor may remark that there are unpumpable contents in the system. In these situations, the tank must be filled up again to cover the steam coils, then reheated and discharged again. After completion discharge of high MP pour point cargoes, it is strongly recommended to wash the back lines using hot freshwater wherever possible. Compressed air is of no use. In case of the planned stop of high MP cargo in freezing weather, keep the cargo recirculating. For the recirculation of shorelines which can freeze, this must be discussed in the preloading meeting between the vessel and the terminal, and thereafter agreed in writing. Failure to do so may result in the vessel being blamed for frozen cargo in unlagged or non-thermal jacketed portions of shorelines.

In order to facilitate faster tank cleaning, the Master should always try to get the Dry Tank Certificate as the tanks are emptied one by one. This is especially important for drying tanks previously loaded with vegetable oils and solidifying cargoes which may jam the impeller and freeze the valves. Furthermore, washing as soon as possible after receiving the tank dry certificate, if the terminal permits it, will prevent some solidification and save time. During discharge, the temperature of the cargo must be at least 10°C (50°F) higher than the pour point of the relevant cargo. If the vessel is discharging high MP cargoes (Table 4.1) a surveyor must be present at stripping/sweeping time. If the discharge of high MP cargoes is stopped for any reason (shore tank changeover/barge changeover/shore leak), the connecting hoses and lines on deck must be immediately drained back into the tank in order to avoid the system from becoming clogged with coagulated cargo. Any remaining onboard (ROB) must be noted on the tank dry certificate and considered by all parties as "unpumpable and unreachable, non-free-flowing sediments/sludge." After the discharging of vegetable

Table 4.1 MP of selected cargoes in °C (°F)

MP of selected cargoes	*°C*	*(°F)*
Coconut oil	14–28	(57.2–82.4)
Lard	33–46	(91.4–114.8)
Hydrogenated corn oil	28–35	(82.4–95)
Hydrogenated rapeseed oil	28–38	(82.4–100.4)
Palm oil	23–50	(73.4–122)
Tallow	35–50	(95–122)
Peanut butter	23–45	(73.4–113)
Cresol	11–35	(51.8–95)
Dichlorobenzene	35	(95)
Fatty alcohol	3–40	(37.47–104)
HMD	41	(105.8)
Naphthalene	80	(176)
o-Nitrochlorobenzene	33	(91.4)
Paraffin wax	55–60	(131–140)
n-Pentane	36	(96.8)
Phenol	41	(105.8)
Polyisobutylene PIB	90	(197.6)
TDI	20	(68)
Trichlorobenzene	15	(59)
Undecyl alcohol	15	(59)
Paraxylene	13	(55.4)

oils/palm oils, to prevent them from drying on the bulkheads, the bulkheads must be kept moist until the tank cleaning process begins. In warm weather the drying process is usually fast; however, in cold and wet climates, this can be much longer and slower. It is important to note that hot cargo in adjacent tanks will make the residues dry extremely fast on the bulkheads. Substances with a reasonable solubility (down to 0.2 %) can be removed with water. Be aware that the solubility of some chemicals increases with temperature. The wash water should be at least 15°C (59°F) higher than the melting point of the cargo. For caustic potash, the crystallising point is 9°C (48.2°F) while for caustic soda it is 12°C–15°C (53.6°F–59°F). As per the IBC Code prewash for substances which have a viscosity equal to or greater than 50 mPa at 20°C (68°F) must be washed with hot water (with the temperature at least 60°C (140°F), unless the properties of such substances make washing less effective:

Some cargo residues have exceedingly high MP which makes them difficult to emulsify. To clean such residues, it may be necessary to use a solvent like toluene. The cleaning of PV vent lines is best done with live steam. Precleaning non-drying fats with hot water is best done after first steaming the tanks. Washing pressures and temperatures must be maintained together and must not be adjusted to compensate for either a loss in temperature by reducing the pressure and/or the number of machines. If a problem is experienced with a loss of heat, then washing may be continued, but the washing time should **NOT** count until the required temperature has been reached again. In the pre-planning stage, a proposed time scale (bar chart) must be formulated, to provide forewarning for the engine room. The Chief Officer must also try to ensure that once pressures and temperatures have been reached, they should be kept as constant as possible, without undue fluctuations. This will avoid major fluctuations and associated problems in the engine room.

The temperature of the cleaning solution must be 20°C (68°F) > MP of the cargo. The temperature of the cleaning solution must be 20°C (68°F) > MP of the cargo. Avoiding rapid changes of temperature especially in cold weather will adversely stress the ship's structure. When cargoes with high wax content are carried, care must be taken that temperatures are not allowed to drop below flocculation/wax precipitation temperature as the wax will not revert back into solution no matter how much the temperature is increased. Hence if the actual loading temperature is different from the specified loading temperature the Master must protest in writing. In cases of extreme temperature difference, the Master may at his discretion, cease all loading operations until the matter is resolved. Always keep the chemical operator informed. When calculating the bunkers required for cargo heating a reserve quantity of bunkers should be included in the total to guard against voyage delays due to adverse weather or unforeseen climatic conditions.

Heating required by MARPOL Annex II

Category Y, high viscosity and solidifying cargoes may require prewash, if not heated. These cargoes do not require a prewash provided the following provisions are complied with:

Category Y cargoes with a melting point less than 15°C (59°F)

The discharge temperature should be at least 5°C (41°F) above the melting point of the product.

> *Example: Benzene with a melting point of 4.5°C (40.1°F) should be discharged with a temperature of at least 9.5°C (49.1°F) to avoid the prewash requirements.*

Category Y cargoes with a melting point above 15°C (59°F)

The discharge temperature must be at least 10°C (50°F) above the melting point of the cargo.

Example: Phenol, with a melting point of 40.9°C (105.6°F) should be discharged with a temperature of at least 51°C (123.8°F) in order to be considered not solidifying.

Category Y cargoes with viscosity <50 mPa at discharge temperature

The Master has to obtain the shipping document with the above information from the shipper when transporting such cargoes so that the products will be heated accordingly so as to avoid the prewash obligation if at all possible.

Temperature checks

When carrying heated cargoes, the following parameters are to be checked regularly:

1. Temperature of the cargo at three levels,
2. Inflow temperature of the heating medium (or steam pressure), and
3. Outflow temperature of the heating medium (or steam pressure).

High-density cargoes

Stowage of a nominated cargo in the ship's cargo tanks must be executed in strict compliance with the vessel's cargo loading manual. When the vessel is instructed to load a high-density cargo (i.e., negative API/density more than 1.0 kg/cm^2), the Master must ensure that the following procedure is adhered to strictly when preparing the load plan: the height of the high-density cargo within the tank should be reduced, so that the pressure exerted at the bottom of the cargo tank by the high-density cargo will not exceed the pressure exerted by the full loaded (98%) cargo of unrestricted loading density i.e.: $h^1d^1 = h^2d^2$, where h^1 = Height of high-density cargo, d^1 = Density of high-density cargo, h^2 = Height of unrestricted loading density cargo when the tank is filled up to 98% and d^2 = Unrestricted loading density. The following formula may be referred to when calculating the filling ratio for a tank:

$$\textit{Filling ratio} = 100 \times \frac{\begin{array}{c}\textit{Maximum density to which the tank} \\ \textit{can be loaded in full}\end{array}}{\textit{Cargo density}}$$

Nevertheless, in case of a vessel is nominated to load a high-density cargo, the vessel's head office will usually consult with the relevant Classification

Society before the vessel is authorised to load the cargo. Further to the above, the following provisions should be adhered to:

- Stress, stability, trim and list,
- Certificate of Fitness (i.e., check the list of chemicals attached to the certificate and confirm that the ship is certified to carry the cargo),
- IBC/BCH Code to be referenced and the carriage requirements of the cargo noted along with IMO classification, etc.,
- Charterer's requirements for the cargo,
- If more than one tank is required, the total volume of the tanks chosen should as near as possible be the same, but not less than, the total volume of cargo, keeping dead space and ullage to a safe minimum,
- Certificate of Class to be cross-checked for ensuring chosen tanks have sufficient strength for high-density cargoes,
- Tank coating compatibility,
- Shipping documents for each cargo to load,
- MSDS for each cargo,
- Antidotes and toxic gas detectors for the cargoes being handled as applicable, and
- Crew familiarity with the cargo to be handled.

Cargo quantities

The cargo quantities must be planned and checked in relation to the voyage orders with regard to:

1. Loading capacity (load calculation),
2. International Load Line Zones in the vessel's trading area,
3. Sheer force and bending moment stresses during the various stages of loading and discharging, and
4. Draft and passage limitations en route, canals and in ports of destination.

Filling limits of cargo tanks

The following provisions are to be accounted for in determining the filling limits of cargo tanks:

1. Load density limit of cargo tanks against cargo density,
2. Density of cargo at maximum voyage temperature or discharge temperature against load temperature,
3. IBC code limit of cargo quantity for ship Type 1 and 2 tanks, and

4. FOSFA restrictions for minimum filling limits (>60% volume in order to consider last cargo) the filling limits of the cargo tanks due to temperature variations and overfill limits shall be complied with as per IBC Code. In addition, vessel shall comply with filling limit restrictions due to load density limits of cargo tanks as per the Certificate of Class and Stability Manual.

Stowage limitations of cargo tank structure and fittings on chemical ships

Attention should be paid to stowage limitations due to cargo tank structures and their fittings, i.e., the specifications of a cargo tank will limit product weights and quantities which can be loaded in that tank. Refer to ship-specific Certificate of Fitness, also the BCH/IBC Codes, for lists of cargoes and tank types suitable for their carriage.

Cargoes generating an electrostatic charge

Static electricity is generated by friction that occurs between different materials during relative motion. Electrostatic charges can then accumulate in materials which are poor conductors of electricity or which are good conductors but are insulated. If two such bodies with accumulated static electricity charges are brought close together, and if the difference of potential is great enough, the accumulated charge will jump between them. The primary concern about static electricity is the possibility of generating an incendive spark within a flammable atmosphere. Inerting a tank can prevent the existence of a flammable gas mixture so that no hazard will exist. Static electricity can be generated due to the passage of a liquid through a hose or pipeline, and turbulence within a tank. In normal circumstances, the charge generated is released instantaneously to earth (the ship's structure) because the liquid conducts it, and design features of cargo tank internals will avoid its build-up. Problems from static electricity are most likely to arise when loading cargoes known as static accumulators, often highly refined petroleum products. It is important, quite apart from cargo quality requirements, to make sure that lines which have been flushed with water have been thoroughly drained and that the bottom of the tank is dry before starting to load a static accumulator cargo. At the initial stage of the loading operation, it is important that the loading rate is limited. Until the bottom longitudinals and tank suction are covered, the loading speed of the liquid in the pipeline should not exceed a linear velocity of 1 metre per second (m/s), which corresponds to the data provided in Table 4.2.

Thereafter, loading may be increased to a maximum pipeline speed of 7 m/s. Experience indicates that hazardous potentials with respect to static electricity do not occur if the velocity is below 7 m/s. However, where

Table 4.2 Pipeline diameter loading rate

200 mm	*115 cubic metres per hour*
150 mm	65 cubic metres per hour
100 mm	30 cubic metres per hour

well-documented experience demonstrates that higher velocities have been safely used, an appropriately higher limit than 7 m/s may be employed. The process of static electricity may happen in a chemical tanker in five different stages:

1. An electrostatic charge is generated in the liquid as it flows turbulently through the loading pipeline into the ship's tank. In most liquids, the charge is released instantaneously to Earth because the liquid conducts it,
2. But in some cases, the charge is accumulated in the liquid because the liquid has a low electrical conductivity. Such liquids are called static accumulators and are found among more highly refined products. An electrostatic field is formed inside the tank,
3. A non-bonded projecting object, or something introduced into the tank, can become a potential electrode or spark promoter, collecting the charge from the liquid,
4. When close enough to an earth the spark promoter instantaneously releases its charge in a spark through the atmosphere of the tank, and
5. Such a spark will certainly have enough energy to ignite a flammable vapour. In chemical tanker operations, a flammable atmosphere may be unavoidable.

Static electricity when loading

Certain materials have the potential to accumulate static electricity charges. This is usually a result of physical activities that have taken place with or around the material. Contact electrification in fluids depends upon the presence of ions or small charged particles. A flowing liquid in a pipe has a certain velocity profile; due to friction at the pipe wall the fluid there will have a lower velocity profile compared to the centre of the pipe. Positive charged particles inside a liquid will be attracted to the grounded metal duct wall and will end up in the liquid film at the pipe wall flowing with lower velocity. The negative particles end up in the inner part of the liquid stream. The moment the liquid flows out of the pipe the positive charge present in the slower moving film at the pipe wall will flow back to the grounded pipe wall. The inner liquid, moving faster and containing a negative charge, cannot flow back and flows off with the liquid into, for instance, a cargo tank. Onboard our vessels we are working with materials and cargoes that have the potential to accumulate static electricity. Naturally, all seafarers are trained on this subject and know how to avoid static electricity climbing to a dangerous level.

All petrochemical products with an electrical conductivity of less than 50 pS/m (= pico Siemens/metre) are so-called non-conductors and are considered to be capable of generating spark hazards. After many studies, it is determined that during handling of a static accumulator cargo, the product may pick up sufficient charge to constitute a hazard. Avoiding sparks is a top priority, but where and how can they originate? In daily life, when walking across a carpet, electrons move from the rug to you. Now you have extra electrons. Touch a doorknob and "ZAP"! The doorknob in this case is a conductor. The electrons move from you to the knob, causing you to receive a minor electric shock. In fact, the same could happen onboard vessels, for example when taking a cargo sample or on the surveyor's request take ullage with a metal ullage tape. If non-conducting equipment is used like a synthetic fibre rope attached to a sampling cage or an ullage tape not properly bounded, the non-conductivity may create a hazardous situation. The sampling cage can be charged by induction if suspended via a non-conductor, like the synthetic rope. A non-conductive rope can also be charged by rapidly slipping throw gloved hands for an appreciable distance. By this, an insulated person can also become charged. Conductive sampling and gauging devices should therefore be used with a conductive lowering device, for example, tape and cable. To avoid a dangerous situation, conductive sampling and gauging devices should be properly bonded to the tank by means of bonding cable or by maintaining continuous metal-to-metal contact between the lowering device and the cargo tank hatch. If it touched the metal of the tank, the tank wall and sampling dip rod are at the same potential. If we bring this theory back to our day-to-day work onboard the vessel, we are able to determine that static electricity by lowering metal devices in cargo tanks or loading or discharging conductive cargoes can be of a great risk. This electrostatic charge will be discharged either to an earthed (grounded) item or between items having differing potentials.

Ten golden rules to avoid static electricity

1. The most important countermeasure to prevent electrostatic hazard is to bind all metal objects together to prevent potential differences,
2. Onboard in chemical tankers it is essential that the initial loading rate for non-conductive cargoes is below the 1 m/s until the liquid

level reaches about 30 cm in the cargo tank. The maximum loading rate should never exceed 7 m/s,

3. Avoid splash filling in the cargo tank,
4. Do not use cleaning agents like Toluene which has a conductivity > 50 pS/m,
5. Cleaning devices like tank wash machines, portable fans and hoses lowered in the cargo tank after discharge must be conductive and bonded to the tank. Hoses for cleaning purposes must be tested for electrical continuity and the resistance should not exceed 6 ohms per metre,
6. The body can store an amount of energy in excess of the ignition energy for hydrocarbons. A body potential of 10–50 kV can be attained by individuals. When handling flammable liquids, one has to take into account that clothing might be a source of ignition. In zone 0, an area in which a flammable gas mixture is continuously present or is present for long periods, and zone 1, an area in which there is likely to be a flammable gas mixture under normal operating conditions, the following precautions must be taken:
   - Avoid synthetic clothing,
   - Do not change clothing in zones 0 and 1 areas, especially in cold areas, and
   - Wear antistatic shoes,
7. A ship/shore bonding cable is not effective as a safety device and may be dangerous and should be avoided. Terminal regulations may require a ship/shore bounding cable to be connected, however, note that maximum resistance to earth is 10 ohms. If such cable is insisted upon, it should first be visually inspected to see, as best possible, that it is mechanically and electrically sound. The connection point for the cable should be well clear of the manifold area,
8. After loading minimum waiting time of 30 minutes must be adhered to before cargo tanks are opened for sampling and ullaging,
9. Radio transmission (300 kHz–30 MHz) radiates significant energy which can, at distances extending to 500 metres from the transmitting antennae, induce an electrical potential in unearthed receivers like for example mast stays or cranes,
   - Transmission on the abovementioned frequency is not allowed during the periods when there is likely to be flammable gas in the region of the transmitting antennae,
   - VHF and AIS and Satellite transmission normally have low energy transmission and do not produce the same source of energy. However, it must be reminded that the VHF and

the AIS must be switched to low power (1 watt) in the port region, and

10. All personnel involved in the handling of static accumulators must be made aware of the risks associated with handling non-conductive cargoes.

Non-conductive or static accumulators

Certain cargoes are known as "non-conductive or static accumulators." These are cargoes with a typical conductivity below 50 pS/m. Table 4.3 presents value and classification for a range of cargoes classified as "Static Accumulators." For a more detailed description of Static Electricity, refer to ISGOTT, Chapter 3. Other cargoes not on the list may also be classified as static accumulators.

Table 4.3 Examples of non-conductive or static accumulators

Product	*Typical conductivity pS/m*	*Classification*
Non-conductive		
Xylene	0.1	Accumulator
Gasoline (straight run)	0.1–1	Accumulator
Diesel (ultra-low sulphur)	0.1–2	Accumulator
Lubricating oil (base)	0.1–1,000[a]	Accumulator
Commercial jet fuel	0.2–50	Accumulator
Toluene	1	Accumulator
Kerosene	1–50	Accumulator
Diesel	1–100[6]	Accumulator
Cyclohexane	<2	Accumulator
Motor gasoline (benzene)	10–300[6]	Accumulator

a Some additives used for performance improvement can increase conductivity significantly.

Chapter 5

Cargo handling operations

GENERAL

The chemical data sheet is a document which, in accordance with the International Maritime Organisation (IMO) Code and usually from the manufacturer of the cargo (the Cargo Information Form), contains necessary information about the properties of the chemical for its safe carriage as cargo. Careful study of the data sheets is essential during the cargo planning stage for various chemical cargoes. This is to ensure due account is taken for the safe stowage and segregation requirement of different chemicals. Every oil and chemical tanker must have a Procedure and Arrangements Manual (P&A Manual) that provides the procedures for complying with MARPOL Annex II. This annex to the MARPOL regulations concerns the carriage of noxious liquid substance (NLS) cargoes by ships at sea. The following sequence outlines a general cycle of operations, with supplementary comments where relevant:

- Preparation for cargo loading,
- Inerting/Purging,
- Loading,
- Transport,
- Preparation for discharge,
- Discharge,
- Ballasting, and
- Tank-cleaning/Gas-freeing.

Outlined below is some useful guidance for handling various noxious liquids. This should be considered general guidance only, as there is considerable variation in the design of cargo containment and cargo handling systems. In all cases, the cargo containment and handling systems will have been designed and constructed in accordance with the requirements of the IBC/BCH codes, the Safety of Life at Sea (SOLAS) convention and the MARPOL 73/78 convention to safely transport and handle the types of chemicals the

DOI: 10.1201/9781003505044-5

ship is certified to carry. However, the required levels of safety in cargo operations can only be achieved if all parts of the system and equipment are maintained in good working order. Similarly, the personnel involved in cargo operations must be fully aware of these instructions and their duties and be thoroughly trained in the correct procedures and handling of the equipment. Before and during all operations involving the cargo, ballast and bunkering systems, the Master must ensure that the precautions required by the vessel's Safety Management System and relevant checklists are fully observed. Each vessel that is certified for the carriage of NLSs in bulk is provided with a ship-specific P&A Manual. All substances permissible for carriage onboard are listed in this manual and are approved for and on behalf of the Flag State government that the vessel is registered under, usually by a Classification Society acting on its behalf. The purpose of the P&A Manual is to identify the arrangements and equipment to enable compliance with Annex II of MARPOL 73/78 and to identify for the benefit of the ship's officers all operational procedures with respect to cargo handling, tank cleaning, slop handling, residue discharging, ballasting and deballasting, which must be strictly followed. If the vessel has been nominated to load cargo that is not listed in the P&A Manual, this should be immediately notified to the vessel's operator. They will then liaise with the Classification Society to determine whether a note of acceptance or a dispensation is possible, which will allow the cargo to be carried onboard.

The details of permissible substances for carriage onboard are detailed in an attachment to the Certificate of Fitness issued by the vessel's Flag State, usually as a delegated responsibility to vessel's Classification Society. The P&A Manual must be updated as necessary to reflect any changes to the vessel's structure, tank coatings etc. Any alterations to the P&A Manual must be Class approved. A list of actions which may be referred to when discussing the cargo plan is provided here: (1) load the vessel so that positive trim is ensured during discharge, preferably without filling ballast in cargo tanks, and particularly not in port. Try to find out the receiver's desired sequence of discharge. Always maintain an eye on hogging and/or sagging! (2) Inter-reactive cargoes must not be placed in neighbouring tanks. Piping systems must be separated by double-blind flanges to prevent the erroneous handling of valves. Check each of the cargoes for cargo compatibility; (3) toxic cargoes must not be placed in neighbouring tanks with edible products (human or cattle). Separate the piping systems by means of double-blind flanges; (4) check with the tank coating manufacturer's list of permissible cargoes for coatings in each tank.

The general rules suggest zinc silicate coatings are resistant to strong solvents (aromatics, alcohols, ketones, etc.); however, zinc silicates are not resistant to caustic soda or alkaline cleaning chemicals. Epoxy coatings are resistant to petroleum products, caustic soda, vegetable oils, wine, seawater and fatty acids and offer limited resistance to alcohols and aromatics. Coal tar epoxy is resistant to seawater, crude oil and petroleum products in general but should not be used for jet fuels or light oils as they tend to be contaminated by bleeding tar. In certain cases, the tank coating manufacturer may provide a limited acceptance for a product (time and/or temperature). Avoid then placing heated products on the other side of the bulkhead. Allow epoxy to weather out properly after carrying solvent cargoes. Never fill a tank with ballast water immediately after emptying methanol from the same tank. Polymerisable products (e.g., styrene and vinyl chloride) should never come into contact with a bulkhead with heated cargoes. The same refers to drying vegetable oils (e.g., linseed oil). Volatile products (such as aromatics, ketones and alcohols) should not be stowed into bulkhead contact with heated cargoes to avoid unnecessary evaporation losses. Cargo tanks are normally inspected and approved prior to loading. This does not necessarily relieve the vessel of responsibility for preventing cross-contamination. The Master carries the responsibility for taking due care of the cargo. To protect the vessel against future claims, it is strongly recommended the Master carries out their own inspections during the loading process. These inspections should be recorded in the deck log.

After cargoes with a strong odour (e.g., fish oil, phenol, octanol, tall oil, turpentine and molasses), the tanks should not be used immediately for odour-sensitive cargoes such as glycols and vegetable oils. Where leaded gasoline products have been discharged, cargoes for human or animal consumption must not be loaded as the next cargo. This includes virgin naphtha feedstock. The reason being that lead compounds may adhere to the bulkheads after several intermediate cargoes, even in coated tanks. Moreover, wine cargo may dissolve lead residues, which are many intermediate cargoes "old." In any case of doubt regarding the purity of the cargo to be loaded, always take cargo samples from the loading manifold upon loading and have them sealed and identified for future reference. In tanks which have contained products with a high boiling point and/or low water solubility (e.g., lubrication oils), there will be minute amounts of cargo left after washing. These tanks will remain unsuitable for "sensitive" cargoes such as methanol.

Always consult the vessel's cargo trim and stability book. There may be restrictions with regard to cargo distribution and stability in hypothetical damaged conditions.

When the same pipe has to be used for several consecutive products, always start with the lighter products before going on to the more viscous products. The pipe may have to be drained and steamed in between processes; therefore, try to arrange an open loop. The most sensitive cargoes may have to be loaded "over the top" through a deck hatch.

It should be borne in mind that every individual ship will have its own characteristics and limitations, especially when handling various chemical cargoes. The Master, and indeed all members of the ship's crew, must be aware of any pertinent cargo/ship information that has been provided and thereafter comply without exception or deviation. A list of reference publications which may be useful when planning cargo is provided at the end of the book.

LOADING OPERATIONS

The loading operation must be carried out in accordance with the cargo operation manual as prepared by the vessel's shipyard and thereafter approved by the Classification Society for the vessel, taking due account of the following factors.

Investigation of cargoes to be loaded

The Chief Officer must investigate the items listed below in planning the loading plan. Accordingly:

1. **Compatibility of cargoes.** Compatibility of cargoes must be checked with the IMO Certificate of Fitness for the Carriage of Dangerous Chemicals in Bulk or the International Certificate of Fitness for the Carriage of Dangerous Chemicals in Bulk,
2. **Properties and hazards of cargoes.** Properties and danger of cargoes must be investigated with reference made to the relevant Operation Manual Data Sheet, Chemical Data Sheet (USCG), Chemical Handling Manual (NKKK), Chemical 800 (NKKK) and/or Material Safety Data Sheets,
3. **Loading plan.** Taking in account the factors listed below, the Chief Officer should make a safe and efficient loading plan before submitting the plan to the Master for approval. The cargo loading plan must include or take account of the following:
   a. Quantity and type (IMO) of cargoes,
   b. Strength and coating compatibility of tanks,
   c. Relative position of toxic cargoes and accommodation spaces,
   d. Segregation of cargoes reacting with heat from heating cargoes,
   e. Necessity of sealing up cargoes with nitrogen,
   f. Restrictions on loading edible oils due to previous cargoes (refer to the FOSFA Banned List),
   g. Applicable load lines according to the Chart of Zones and Seasonal Areas,
   h. The deadweight corresponding to the above and available cargo capacity of the vessel. The cargo quantity to be loaded should be planned so that 98% of the tank capacity may be utilised at the maximum temperature expected while loading/discharging

and while at sea. In the case of cargoes which require heating, allowance should be made for expansion due to heating,

i. Conditions of ports and trade route, and
j. Available depths of water, permissible draft or under keel allowance, specific gravity of the seawater and tides at ports, harbours, berths, straits, rivers and canals.

4. Filling limits of cargo tanks on their designed specific gravity, i.e.:

$$Maximum\ filling\ ratio\left(\frac{O}{o}full\right) = \frac{DSG}{SG} \times 100$$

where DSG is the designed specific gravity of the cargo tanks; and SG is the specific gravity of the cargoes to be loaded.

5. **Filling limits of cargo tanks on their sloshing strength.** In general, partial loading between 20% and 80% of filling ratio should not be adopted to avoid an excessive sloshing load on the tank structure. However, if partial loading is involved in the intended plan, cargoes should be loaded to the extent of a safe loading percentage as determined after referring to the sloshing calculation shown in the vessel's loading operation manual,
6. The loading operation manual must be referred to when calculating the vessel's trim/stability parameters,
7. The ship's survival stability in damaged condition, as required by IBC Code, must be consulted,
8. Cargo suitability/resistance table for coated cargo tanks,
9. **Hull strength.** On the basis of the loading operations manual provided for each vessel, calculations should be made of the stresses to be imposed on each part of the hull after loading, to ensure that such stresses are within the permissible range of safety. Stress should be considered so that the hull stress is kept within the permissible range not only after the finish of cargo loading but also during the loading and discharging operations. Draft and trim should also be appropriate. The result of the calculations should be recorded and retained on board.
10. **Loading rate.** The maximum loading rate acceptable for the vessel or for each tank should be determined by the following factors, as well as the current condition of the ship's facilities and the experience and capabilities of the crew members concerned, with reference to the loading manual:
    a. Number of tanks to be loaded,
    b. Loading rate corresponding to the maximum flow rate of 7 m/sec for cargo to pass through the ship's pipeline (including the manifold),
    c. Loading rate corresponding to the maximum flow rate of 38 m/sec for the gas inside tanks to pass through the vent line,

Preparations before arrival at loading port

Check and carry out operational tests of cargo handling equipment and instruments:

1. **Checks of gas leaks through openings on deck and vent systems.** In calm weather conditions while at sea, gas leaks from openings, such as tank cleaning holes, tank hatches, ullage holes and any other opening, must be checked for leaks, if any, and must be stopped with the necessary measures taken, for example, tightening the relevant bolts and nuts and replacing packing,
2. **Operational test of tank level gauges.** An operational test should be made on the level gauges to check the matching of the readings shown locally and remotely in the cargo control room (CCR) to the normal function of high/overfill and low-level alarms, and the results should be recorded, and
3. **Check of heating coils.** Heating coils for cargo heating should be pressure tested by steam or compressed air of more than 7 kg/cm on a regular basis, and the results of tests should be recorded. The leaking steam coils should be blanked off and this information should be maintained in the records and displayed in the CCR.

Loading procedures

Meeting with shore responsible person prior to the commencement of cargo operation

The Chief Officer must discuss the following matters with the shore responsible person on the basis of the loading plan:

1. Information on safety, including the terminal safety regulations and safety checklist. The Chief Officer must fill out and comply with the International Safety Guide for Oil Tankers and Terminals (ISGOTT) safety check list, including the chemical requirements, even if the terminal does not have one,
2. Designation of smoking places,
3. Restrictions on the use of fire and cooking devices in the galley,
4. Items which require the posting of notices,
5. When operations which involve hot work or other repair work are conducted on board the vessel or in the shore facilities, such place and method,

6. Means and methods of communication between the vessel and terminal,
7. Emergency measures (emergency shut-down procedures),
8. Designation of access doors to the accommodation spaces,
9. Types and expected quantities of cargoes and loading sequence. Loading must be commenced only after complete agreement on the cargo to be handled is reached between the Chief Officer and the terminal responsible person. If there are discrepancies between the instructions given by the charterer or the Company and requirements made by the terminal, the charterer or the Company should be contacted to clarify the situation before the commencement of the operation,
10. Matters concerning slops,
11. Loading rate upon the commencement of loading, maximum allowable loading rate, loading rate for topping off and procedures to change over to a different kind of cargo,
12. Confirmation of whether cargo loading is ordered to stop by the shore or ship's responsible person, the stand-by period required for ordinary stops, and the amount of cargo which flows in after the order to stop loading,
13. Discharge or loading of ballast water, its quantity and required time,
14. Details or cargo (e.g., temperature, water content, properties, precautions in handling and loading plan),
15. Matters concerning the ship's facilities (e.g., the pump capacity, and the present condition of the inert gas system, tanks and pipes),
16. Matters concerning shore facilities (movable scope of the loading arm, connecting/disconnecting method, present state of pipes and tanks),
17. Matters concerning the procedures to measure the quantity of cargo, taking samples, take temperatures and take water cuts,
18. Matters concerning restrictions, for example, on the height of the ship's manifold above the water (height of the bow fair-lead), draft and trim,
19. Schedules and quantities of bunker oil and water to be supplied, and
20. Additional matters which require confirmation concerning cargo work and safety.

Inspection of cargo tanks before loading (dry inspection)

The duty officer, under the direction of the Chief Officer, must check and confirm that the cargo tanks have dried in the presence of the terminal responsible person or cargo surveyor. After the inspection, a tank dry certificate must be prepared and signed by the terminal responsible person or cargo surveyor.

Procedures when commencing loading operations

The Chief Officer must confirm before the commencement of cargo work that the cargo system (valve handling) is correctly lined on the basis of the loading plan. The Chief Officer must attend in person the commencement and finishing of the loading operation. The position (open or closed) of valves should be checked, and the cargo loading system should be set up by observing the closure of all valves (hydraulically driven or hand-operated) on the cargo lines. This will range from those operated at the console in the CCR at local stands in the pump room and on the upper deck to hand-operated valves. In the case of valves which have been kept open, in consideration of thermal expansion inside the pipes, such valve positions should also be confirmed. The cargo line valves for loading should be opened in sequence, from valves close to the manifold gate valve towards the cargo tank to be loaded via intermediate valves, including jumping valves, drop valves and bottom-line bypass valves. The closure of the deck master valves and other valves leading to cargo lines which are not used should be confirmed. The suction valve for one tank among those to be loaded should be opened. The manifold gate valves should be opened lastly after the line-up for loading is completed and the approval of the terminal is obtained. The Chief Officer must check in person the open or closed position of the valves and fitting conditions of blank flanges and post crew members at specified locations in order to make the vessel ready to act in response to the request of the shore to commence loading. Cargo should be loaded at a minimum rate upon the commencement of loading. After checking that no cargo is leaking, no excessive stress is imposed throughout the loading system, including the pipeline and manifold, and cargo is being properly transferred, the loading rate must be gradually increased to a maximum agreed between ship and shore.

When commencing the cargo loading procedure, the Chief Officer must pay attention to the number of cargo tanks to be loaded at the same time, with the motto "**SAFETY FIRST**" in mind, to avoid excessive pressure on the relevant cargo lines and valves. The Chief Officer must station a crew member who is able to communicate with terminal personnel, in the vicinity of the manifold to monitor the situation upon the commencement of loading operations and ensure that he checks and reports the inflow of cargo, manifold pressure and temperature and cargo leaks from the connections so as to provide against emergency situations. When cargo handling operations are commenced or resumed on board a vessel with trim, it must be taken care to monitor differences in liquid levels in cargo tanks to avoid overflow. When loading is suspended temporarily, the suction valves for tanks which contain cargo must be securely closed to prevent overflows as a result of shifting of cargo from tank to tank.

Discharge of ballast water for arrival

The discharge of water ballast must always be made, in any case, under sufficient surveillance by complying with the procedure specified in the ship's standard operating procedure (SOP) for handling of water ballast, to prevent releases of pollution out to sea. Even in the case of deballasting from a segregated ballast tank, cargo may mix into water ballast through holes of the pipelines or bulkheads. The sea surface in the vicinity of the discharge opening should therefore be monitored at regular intervals during deballasting operations. At this time, attention must be paid to the difference in the head height among the cargo tanks, ballast tanks and seawater.

Monitoring of ullage

The ullage of every tank must be monitored during cargo handling operations to prevent overflows. And efforts must be made to detect unusual conditions, including leaks from valves, at an early stage. It must be monitored at least once every hour, and the loaded cargo quantity of each tank should be calculated to see whether the cargo work has progressed according to the loading plan and such figures to be recorded. When the ullage space has decreased, it requires continuous monitoring. On board vessels where ullage is monitored with remote level indicators in the CCR, the readings of the indicators must be checked to see that they show correct figures before crew members become constrained by time at the final stage of loading cargo. In the case of vessels equipped with float type level gauges, they should be checked to see that they properly function without sticking. When the completion of loading is drawing near, every precaution should be taken to prevent overflows by taking proper measures, for example, reducing the loading rate. For this reason, communication with the terminal responsible person must be ensured so that sufficient care is taken in the operation of the valves and pumps on the terminal side. Close attention must be paid, during cargo handling operations, to the ullage of tanks which are not under loading, to prevent unforeseen accidents caused by leaks of valves and pipelines. Even ullaging devices for the tanks which are not under loading operation must not be stopped until all operations are completed.

Topping off

In the final stage, loading operations should be adjusted to prevent overflows and to take into consideration the allocated manpower, so that tanks may be topped off one by one by avoiding simultaneous operations at more than one tank. In addition, with the decrease in the number of tanks being loaded, the rising speed of the cargo level increases, which requires slowing down of the loading rate in sequence. A centre tank in the midships section should preferably be selected as the final tank to be topped off, so that loading in the final stage may not affect the trim or heel of the vessel. Before topping off the final tank, the approximate or final gravity should be obtained from the terminal to calculate the loaded cargo quantity using drafts, cargo temperatures and ullages to adjust the ullage of the final tank. When the final tank is to be topped off, the Chief Officer should ensure that the loading rate is reduced to a minimum and that the terminal is on standby, ready to stop loading at any time. The Chief Officer should also ensure that crew members report the ship's draft and the ullage of the relevant tank, so as to give a stop order so that the vessel may load the intended cargo quantity. When the expected ullage for the final tank is small, an empty tank or a tank with a large ullage should be identified so that it can be used in an emergency case by opening the suction valve of the very tank to receive cargo oil, thereby preventing overflows. In order to prevent damage to the cargo transfer system between the shore and the vessel due to pressure surge effect, the tank suction valves must not be shut off against the cargo pressure before the notification that the cargo transfer has finished, is received from the terminal, even if the cargo has been loaded in excess of the expected quantity. Even after the loading is finished and there is no inflow of cargo, the ullage should be monitored until the rise of the cargo level settles, since the cargo contained in the cargo lines on deck, drains into the final tank.

Completion of the cargo loading operation

When finishing cargo loading operations, the ship's valves should be shut only after the shore valves have been confirmed to have been shut. At the end of the loading operation, the disconnection of cargo hoses or arms must be carried out in the presence of the Chief Officer. Close attention should be paid to prevent cargo spilling upon disconnection by, for example, opening drain valves to drain remaining cargo in the hoses or arms before disconnecting them. The manifold gate valves should be closed, and the cargo remaining in deck lines should be cleared into tanks, as far as possible, by opening air inlet valves or drain valves. Each valve should be closed thereafter. It should be ensured that tank openings such as ullage holes and the vent riser valves are closed.

Cargo measurement and water cut

On completion of the loading operation, the duty officer, together with the cargo surveyor, under the direction of the Chief Officer, should measure the cargo quantity in the loaded tanks, take water cuts and temperatures and make an ullage report. A calculation of the loaded quantity of each tank should be performed by using the cargo temperature and volume reduction of each tank, with the results entered into the official ullage report. Tank openings, including ullage holes, should be opened only when they are used to take measurements and temperature, and to be closed immediately thereafter. The measurement of the cargo quantity should be taken as follows:

1. After the completion of cargo loading, measurements should be taken of all tanks where cargo has been loaded. The gauging equipment should be in good working order.
2. Onboard vessels equipped with float gauges or radar echo type gauges, attention should be paid to ullage readings; if they involve errors, such errors should be recorded in the "Table of comparison between actual ullage and float gauge" and cargo work should be carried out with the errors in mind. The gauging devices should be adjusted to eliminate errors after the completion of cargo work.
3. The measurement may be taken by sounding or ullaging and the cargo quantity calculated by using tank tables provided on board the vessel. Measurements should be taken up to an accuracy of 0.5 cm. In most terminals, closed loading and sampling are the common routines, and these should be complied with as per the requirements of the terminal/port/state.
4. When calculating cargo quantities, corrections for trim and heel should be allowed.
5. Finally, it is the responsibility of the Chief Officer to ensure that cargo measurements and sampling are carried out accurately. He should ensure correct tables are used for the right type of cargo. Care should be taken for obtaining the correct cargo density.

Cargo temperatures should be taken as follows:

1. Cargo temperatures should be taken in all tanks loaded, as a rule, unless specifically instructed otherwise.
2. Fixed thermometers should be checked for errors prior to use, and such errors, if any, shall precisely be grasped.
3. When there are two temperature-detecting points in tank, the mean of each reading should be adopted. Average of the calculation should be used.
4. If ship/shore difference of cargo figures of more than 0.3% is found and no reason can be attributed, a note of protest should be logged.

Samples should be taken as follows:

1. When a person concerned with the terminal takes samples, he should do so only after obtaining the permission of the Chief Officer.
2. In principle, samples taken by the cargo surveyor should be kept on board. If this is not practicable, the Chief Officer shall take samples of cargo and keep them on board. The samples taken by the Chief Officer should be signed by the shipper's representative and then sealed also.
3. Samples shall be taken from the sampling cock on the ship's manifold by the shippers at the commencement of loading and when the liquid level of tanks become one foot from bottom. When obtaining a sample of a refined product, a clean sampling device must be used.
4. Samples should be retained in the designated place on board. They should be labelled with the kind of cargo and date of sampling, and care taken to protect them from damage; use a specially designed sample locker to store the samples. The samples should be retained onboard for at least two years.

DISCHARGING OPERATION

Discharging operations must be carried out in accordance with the cargo operation manual as approved by Class for the specific ship, taking into consideration the following factors.

Discharging plan

For making a discharging plan, the Chief Officer must check the hazard ratings of cargoes with the health hazard data described in the United States Coast Guard (USCG) Chemical Data Guide for Bulk Shipment by Water and the material safety data sheets. Furthermore, the Chief Officer must consider the following factors to ensure the safety and efficiency of the discharging operation. The plan must be submitted to, and approved by, the Master:

- **Discharging sequence**, discharging time and pipelines and pumps to be used,
- **Checking the conditions of the vessel** (changes in draft, trim and stress on the hull) in each stage of the discharging operation and make a ballasting plan,

- **Hull strength.** The hull strength should be within permissible limits at any time during discharging operations and, if the vessel is to discharge at two or more ports, during navigation between such discharging ports,
- **Berth conditions.** Berth type, condition of shore reception tanks, clear height, draft restrictions, reception of slops and ballast water, restrictions on pressure and flow rate, line flushing after the completion of cargo discharge etc., and
- **Discharging rate.** Permissible capacity of the shore facility to receive cargo (including permissible rate for hoses or arms), capacity of the pumps of the vessel, discharging rate corresponding to the maximum flow rate in the cargo lines of the vessel etc.

Preparations before entering discharging port

When preparing to enter the port of discharge, the following inspections and operational checks of the cargo equipment and instruments must be carried out:

1. **Pressure tests on discharging cargo lines for checking leaks.** A pressure test should be conducted by pumping cargo from a convenient tank and by applying about 150% of the working pressure with a pump on cargo lines to check for leaks from joints and valves of each line in particular and the results recorded. Any defective parts, if any, must be repaired or other necessary remedial action be taken. After the pressure test, cargo contained in the pipelines should be cleared into tanks in order to prevent damage to valves and lines due to expansion of cargo or clogging of lines due to solidification of cargo in the case of a heated cargo.
2. **Leaks check for cargo pumps.** Seal of each cargo pump shall be cheeked for leaks, and extent and kind of leaks, if any, shall be confirmed and recorded. Any leaking parts must be repaired.
3. **Checks of gas leaks through openings on deck and vent system.** In calm weather conditions while at sea, gas leaks from openings such as tank cleaning holes, tank hatches, ullage holes and peep holes must be checked, and leaks, if any, must be provided with necessary measures to rectify the situation, for example, tightening the bolts and nuts and replacing gasket.
4. **Operational test of tank level gauges.** An operational test should be made on the level gauges to check the matching of the readings shown locally and remotely in the CCR, the normal function of high/overfill and low-level alarms should be checked and the results to be recorded.

Discharging procedures

Before discharging of the cargo commences, a meeting between the senior ship's officers and the onshore responsible person must take place to discuss the following matters in relation to the conduct of the cargo discharge plan:

- Information on safety, including the terminal safety regulations and the ship/shore safety checklist,
- Designation of smoking places,
- Designation of access doors to the accommodation space,
- Restrictions on the use of fire and cooking devices in the galley,
- Items which require the posting of notices,
- When operations which involve hot work or other repair work are conducted on board the vessel or in the shore facilities, as the place and method of such operations,
- Means and methods of communication between the vessel and the terminal,
- Emergency measures (emergency shut-down procedures)
- Discharging sequence, expected discharging quantity and the expected tanks to be discharged,
- Discharging rate upon commencement of cargo work, maximum allowable discharging rate, manifold pressure and procedures to change over to a different kind of cargo and to finish discharging,
- Method to check tanks which are to be dried up,
- Loading of ballast water, its quantity and required time, method of ballasting and restrictions on deballasting,
- Matters concerning the ship's facilities (e.g., the pump capacity, and the present state of tanks and pipes),
- Matters concerning restrictions, for example, on the height of the ship's manifold above the water (height of the bow fair-lead), draft and trim,
- Matters concerning shore facilities (movable scope of the cargo arm, connecting and disconnecting method, present state of pipes and tanks, special circumstances and requirements of the berth),
- Matters concerning the procedures to measure the quantity of cargo, take samples, take temperatures and water cuts,
- Schedule and quantities of oil and water to be supplied,
- Additional matters which require confirmation concerning cargo work and safety, and
- Legislation or voluntary requirements regarding mid seawater ballast exchange to avoid the discharge of harmful bacteria in territorial waters to be complied with.

Cargo measurement, etc.

Before the commencement of cargo discharge, the duty officer, under the direction of the Chief Officer, should measure the cargo quantity in the presence of the terminal responsible person or the cargo surveyor, take temperatures and make an ullage report. The duty officer should calculate the cargo quantity of each tank by using the oil temperature and volume reduction of each tank and enter the results in the ullage report. Tank openings, including ullage holes, should be opened only when they are used to take measurement and temperature, and closed immediately thereafter.

Procedures when commencing discharging operations

The Chief Officer must confirm before the commencement of cargo work that the cargo system (including valve handling) is correctly lined up on the basis of the discharging plan. The Chief Officer must attend in person upon the commencement and completion of the discharging operation. The position (open or closed) of valves should be checked, and the cargo discharging system set up in accordance with the vessel's SOP. The closure of all the valves (hydraulically driven or hand-operated) on the cargo lines, ranging from those operated at the console in the CCR, at local stands in the pump room and on the upper deck, to hand-operated valves, as a rule, should be checked. In the case of valves which have been kept open in consideration of thermal expansion inside the pipes to be used in the opened position, such valve positions should be confirmed. The above checking operation may be carried out for the purpose of improving the efficiency of cargo work when sufficient time is available, for example, at anchorage before berthing. The cargo pipelines should be set for discharging on the basis of the discharging plan; the cargo pumps should be bled of air. The Chief Officer must check in person the open or closed status positions of the valves and fitting conditions of blank flanges of unused manifold and shall station crew members at specified locations in order to make the vessel ready to respond to the onshore request to commence discharging. After testing and confirming that the turning of the cargo pumps is in good order, cargo discharge should start at a minimum rate. This must be done in the presence of the terminal responsible person. After confirming that no leak is observed or excessive pressure is applied on the whole cargo transfer system, including the pipeline and manifold, and that cargo transfer is being carried out properly both on board the vessel and ashore, the discharge rate can be increased gradually to its maximum permissible limit (flow speed or manifold pressure).

Upon the commencement of discharging operations, a lookout must be stationed to monitor the vicinity of the manifold. The lookout should also monitor the free outflow of cargo, manifold pressure and communicate any cargo leaks from the connections to prevent emergency situations. At the same time, a person must be assigned to the pump room and another on deck to monitor the upper deck and sea surface for indications of cargo spills. Where the vessel does not have a dedicated cargo pump room, the operation may be monitored either from the CCR or locally on deck. The mechanical seals and bearings of the cargo pumps must be checked, upon the start of the operation and with any changes in loads. Even if the vessel is equipped with remote temperature indicators, these must be checked regularly. For pumps, it is critical that the manufacturer's instructions are followed. When cargo handling operations are commenced or resumed onboard a vessel with trim, discharging must be carried out with any differences in the tank levels taken into consideration to avoid overflows. When discharging is suspended temporarily, the suction valves for the tanks which contain cargo must be securely closed to prevent overflow as a result of cargo being shifted from tank to tank. Until the maximum cargo discharge rate is attained, all deck crew must be engaged, under the direction of the Chief Officer, in checking for leaks and monitoring for unusual conditions in the operation of the cargo transfer system.

Monitoring of ullage

During cargo handling operations, the ullage of all tanks must be monitored. The quantity of discharged cargo, and the balance remaining onboard (ROB), must be calculated to determine whether the discharging operation is proceeding according to the discharging plan. These checks must be made at least once per hour with the figures recorded in the deck logbook. Once the cargo level in the tank drops below the predetermined level, the ullage should be monitored continuously. Where ullage is monitored with remote level indicators in the CCR, the readings of the indicator must be checked manually to prove they show correct figures before the crew becomes time-constrained at the final stage of discharging the cargo. Vessels equipped with float type level gauges should be checked to confirm that they are properly functioning. Special attention must be paid during cargo handling operations to ullage in tanks in which no cargo is handled to prevent unforeseen accidents caused by leaking valves and pipelines. Ullaging devices for those tanks in which no cargo is handled should not be stopped until all cargo handling operations have ceased.

Heating

When the heating of cargo (Figure 5.1) is specifically requested by the shipper as a measure to improve the efficiency of cargo discharge and to prevent cargo outturn shortage, the cargo must be heated as instructed. However, if a request is made for heating in excess of the permissible temperature of

Figure 5.1 Heating arrangements of a chemical tanker.

the ship's facilities, such requests must be declined. Under no circumstances should the integrity of the vessel and her crew be unduly undermined.

Stripping

In order to minimise transportation losses of cargo oil, the vessel should be trimmed by the stem as far as possible within a safe range with stripping performed completely.

Completion of cargo discharge

When the discharging operation is complete, the valves on the ship's side must be closed, after which the terminal must be notified. On completion of the discharging operation, the disconnection of cargo hoses or arms must be carried out in the presence of the terminal responsible person and the Chief Officer. Special care should be taken to prevent oil from spilling upon disconnection, for example, by opening the drain valves to drain off any remaining cargo in the hose or aims before disconnection. Once the cargo has been fully discharged, the residual cargo in the deck pipelines (including small lines) should be drained into a tank after closing the manifold gate valves; thereafter, such valves should be closed. Any tank openings, including the ullage holes, must be closed and secured.

Check of cargo tanks after cargo discharge

Upon completion of cargo discharge, the duty officer must check, under the direction of the Chief Officer, whether the cargo tanks have dried up. This is usually done in the presence of the terminal responsible person or the cargo surveyor. Thereafter, the signature of the terminal responsible person and/or cargo surveyor should be obtained on the tank dry certificate. In some locales, the certificate is called the Empty Tank Certificate. Either document is acceptable provided it is signed by either/or the terminal responsible person or cargo surveyor.

SHIP-TO-SHIP TRANSFER OPERATIONS

When ship-to-ship transfer operations are carried out at sea, the procedures listed below must be observed in addition to those items described in the document *Ship-to-Ship Transfers (Petroleum)*, as published by International Chamber of Shipping (ICS) and the Oil Companies International Marine Forum (OCIMF).

Procedures for ship-to-ship transfer operations

1. Main engine and auxiliary machineries must be on standby for immediate use.
2. Adequate supply of fenders shall be placed between both ships.
3. Cargo hoses connected to the ship's manifold shall be ready for quick disconnection in emergencies.
4. Mooring of both ships shall be arranged so that flammable and/or toxic vapours vented from either ship shall not enter into the ship's accommodation spaces and/or engine room or shall not accumulate in dangerous concentrations around the deck.
5. When handling cargoes which generate static electricity, bonded cargo hoses shall be used, otherwise, both hose flanges – including intermediates – shall be bonded externally.
6. The ship commanding the operation shall be decided in advance.
7. P&A provider should be advised prior to the operation with all information needed for insurance cover.
8. All permissions and authorisations must be obtained.

SHIP-TO-BARGE TRANSFER OPERATION

For ship-to-barge transfer operations, the following precautions must be observed in addition to those outlined above.

Procedures for ship-to-barge transfer operations

1. The operation should be carried out in favourable weather conditions.
2. All persons engaged in the transfer operation must be familiar with the nature and hazards of the cargoes to be transferred and all safety precautions are to be observed.
3. The mooring of the ship and barge shall be done in such a manner that the ship and barge can be quickly released from each other in an emergency.
4. All operations must be suspended immediately whenever safety is compromised, no matter how minor the infraction may appear.
5. On completion of the transfer operation, the barge shall be removed from alongside as soon as possible after the completion of loading or discharging.

Chapter 6

General precautions on cargo handling operations

GENERAL

In terms of general precautions and preparations for cargo handling, there are a number of key tasks that must be completed to ensure cargo handling is performed both safely and efficiently. In order to prevent cargo spills, the deck scuppers must be plugged in completely by installing expandable rubber scupper plugs. Rainwater accumulated on deck in rainy weather should be frequently drained with permission from the terminal responsible person after confirming no cargo liquids are mixed. The spill tank should be cleaned with any dirt, rust, water and oil/cargo removed. It should also be ensured that all valves and plugs are closed. Before connecting the hoses or arms, oil cargo spill response materials such as sawdust and rags must be arranged in the vicinity of the manifold in order to respond to oil/cargo spills. When hoses or arms are connected or disconnected, appropriate mats must be laid out on the deck in the vicinity of the manifold to prevent the generation of sparks by friction on the deck by the accidental falling of the tools. All manifold flanges which are not connected with cargo hoses and arms should be closed completely with blank flanges fully bolted. It should be ensured that the drain valves and air cocks of the accessory lines connected to the main lines at the manifold are closed before the commencement of cargo handling operations. The manifold flanges which are to be connected should be cleaned of rust or dirt and new gaskets provided. Firefighting facilities must be prepared as per the information provided in the cargo data sheets. Moreover, before the commencement of cargo handling operations, at least two fire hoses fitted with multi-purpose nozzles and connected to fire hydrants should be arranged for immediate use each in the vicinity of the manifold and at the entrance to the pump room. Water should, as a rule, be introduced into the fire mains.

When it is inappropriate to do so owing to cold weather or other conditions (e.g., specific terminal regulations), the fire pump must be arranged for immediate operation. Within the vicinity of the manifold, at least two portable fire extinguishers must be provided. International shore

DOI: 10.1201/9781003505044-6

connections should be provided for immediate use and its storage place should be clearly marked. It must be borne in mind that for some chemicals, water cannot be used for the firefighting so the vessel's fixed firefighting means must be in readiness at all times during cargo loading, discharging, transfer, tank cleaning and gas freeing operations. To provide for emergency situations while mooring, a firewire with an eye at the end must be suspended through a chock on the offshore bow and quarter of the ship. The firewires must be adjusted according to the change in the freeboard and trim so that the eye may be positioned about 1 meter above the water level at all times or as per the terminal requirements. The size of the wire should be as per Oil Companies International Marine Forum's (OCIMF) requirements. When a derrick or crane is to be used to lift up a hose, etc., at a single buoy mooring, it should be planned and prepared to ensure that it endures the load, by inspecting and providing maintenance service on the rotating parts and accessory equipment, including ropes and blocks to see that there is no unusual condition. Gears (cargo fall etc.), in particular, should be removed from kinks and twists completely. When operating a derrick or crane, the command system should be clearly understood by crew members who operate the winch. The orders and signals given by the signaller should be faithfully followed, and abrupt operations which may give shocks to the boom must be avoided. The signaller must take care to ensure that no person is immediately below the derrick or crane or inside the bight formed by blocks and rollers. Before arrival port, earth tests should be conducted and leaking points, if any, be serviced. Attention must be paid to the condition of glass cases for lighting and sheaths of cable leading to masts and posts, and damaged parts, if any, be repaired. All chemical tankers of more than 20,000 DWT must be fitted with emergency towing equipment as per SOLAS. These must be maintained in good condition at all times. Last of all, the emergency shutdown system must be tested every 15 days and recorded accordingly.

Safety measures during cargo handling operation

Watch procedures during cargo handling operations

During cargo handling operations, a sufficient number of crew members must be available onboard to secure the smooth progress of the operations and their safety. The arrangement to keep cargo watches, including those upon the commencement of loading, commencement of handling water ballast and completion of cargo work, should be fixed in terms of the allocation of persons to stations and the roster arrangement and entered into a watch arrangement schedule and posted in the CCR. The duty officer should post at least one crew member in the vicinity of the manifold to provide for emergency situations as well as to monitor the vessel and the surrounding area for oil/cargo spills from the vessel or drifting oil from other sources.

Watch procedures and safety checks at the terminal

The duty officer and crew should constantly check that the cargo handling equipment is functioning properly and confirm the cargo loading operation is in progress at the specified capacity and according to plan; the vessel is properly moored and the accommodation ladder is in proper condition; cargo is not leaking and the water around the vessel is clean; there is no unusual condition with the cargo levels in tanks which are not being loaded; no hydrocarbons or other toxic gases/vapours can accumulate in dangerous concentrations around the vessel; the emission of smoke from the funnel is normal; no vessels other than those authorised to can/are approaching the vessel; no persons other than those authorised to are onboard the vessel; no hazardous work is conducted in places other than those authorised by the Master; the access doors and scuttles of the accommodation space are kept closed; proper lighting is provided at night; necessary signals are displayed or lit; no hydrocarbons or other toxic gases/vapours are present in dangerous concentrations in the accommodation spaces and the machinery spaces; necessary notices are properly posted; warning boards are posted at the gangway (e.g., "No open light," "No smoking," "Dangerous cargoes being handled" and "No boarding without authorisation"); in addition to any relevant cargo data sheets; designation of smoking places and other warnings to attract attention; weather and sea conditions are normal; the cargo hoses are properly connected and maintained; the firewires are properly suspended; and, the requirements of the terminal for draft, trim and other matters are fully complied with. When the terminal responsible person checks safety in accordance with the form specified by the terminal, the duty officer must render cooperation in the check. No cargo handling operation should be commenced before safety is confirmed.

Connection of cargo hoses or arms

The connection of cargo hoses or arms must be carried out in the presence of the Chief Officer or their delegated deputy. The deck officer must inspect the hoses or arms prior to connecting them. This is to confirm they are free from defects. The Master may refuse the use of any hose or arm which appears defective. The gaskets or "O" rings at the connections must be clean and proper. Before lifting up cargo hoses by the ship's derrick/crane, it must be confirmed that the total weight of the hose and its accessories is within the safe working load of the derrick/crane. Attention must be paid when lifting up a hose so that no excessive bending or twisting is given to the hose. The hose must be suspended at a maximum of 3 metres (9.8 feet) horizontal and

15 metres (49 feet) vertical intervals. The hose must be protected by providing materials to reduce friction at points where it comes in contact with metal parts, including the hull, wire or chain. It must be ensured that the hoses are safe against changes in the ship's trim, draft and the tidal level. The manifold flange should be compatible with the hose flange, and both should be securely attached to the hose.

Furthermore, caution must be taken when connecting the cargo arms. Accordingly, the vessel must be positioned when moored so as to place her manifold in position, in relation to the cargo arms, in order to utilise the movable scope to its full advantage. It must be checked whether the manifold is able to support the load imposed by the weight of the arms and the oil contained in them. If there is any doubt as to its capability, proper saddles must be provided. When more than two arms are connected, they must not be crossed. The swivel joints of the arms should be checked for normal operation and the total absence of leaks. The movable scope of the arm must be checked, and the vessel controlled to enable the manifold to be positioned correctly. If necessary, consideration should be given to adjusting the loading and ballasting operations in accordance with the vessel's trim and draft and prevailing tidal levels. Extra caution must be taken to prevent cargo spills when connecting or disconnecting the cargo hoses or arms.

Signals and lights displayed during cargo work

When the vessel is not gas-free, she must fly a red flag (international code flag "B") (Figure 6.1) during daytime hours and illuminate a red light during hours of darkness. When operating in Japanese waters subject to the "Maritime Traffic Safety Law," the vessel must display a first substitute followed by a red flag ("B") in a vertical line instead of the standard daytime procedure and illuminate a statutory red flashing light during night hours. In ports outside Japan, vessels must display the standard red flag ("B") during daytime and illuminate a red light during the hours of darkness unless other local regulations apply. In such situations, the local requirements supersede those outlined above.

Suspension of cargo handling operations

Sometimes it may be necessary to suspend cargo handling operations. This could be in response to particularly inclement weather; when the fire has broken out on or near the vessel; during violent thunderstorms where there is a sustained risk of lightning; when another vessel has collided with or is likely to collide with the vessel; whenever there is an accumulation of hydrocarbon gas/vapours or other toxic gas vapours in a dangerous concentration around the vessel; when other vessels (except small boats) come alongside or leave the vessel; when the vessel's mooring arrangement is threatened

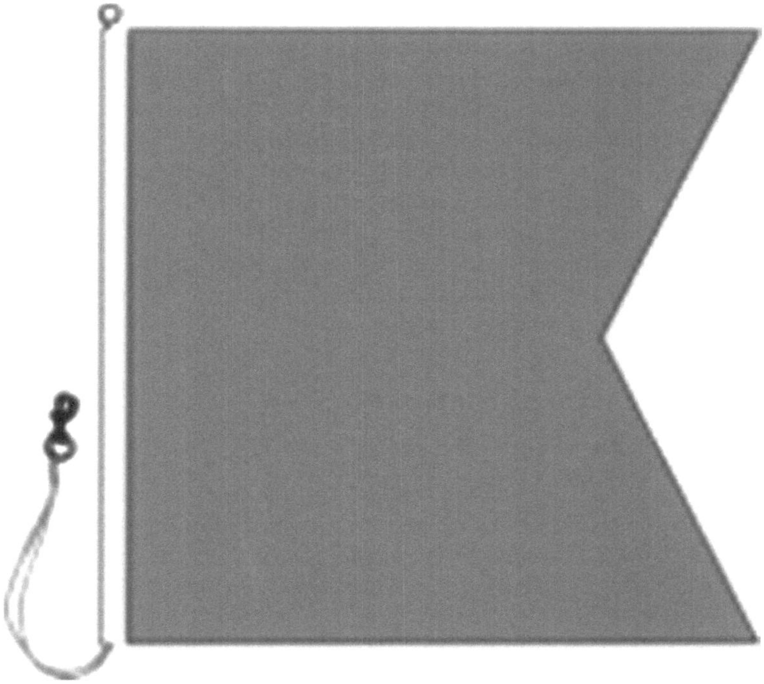

Figure 6.1 International hazard flag "B".

by deteriorating weather or sea conditions; when required by the terminal; when cargo/oil is found around the vessel; whenever a situation threatens to create a source of fire or otherwise compromise safety; and/or when the vessel has broken away from the jetty during cargo transfer. In one of the cases outlined above, the terminal must be requested to disconnect the cargo hoses or arms and implement an emergency response status.

Measures to be taken in an emergency situation

When, during a cargo handling operation, any one of the emergency situations listed above is likely to be or has been encountered, the duty officer must stop the cargo work immediately. They must notify all parties concerned, including the terminal, and obtain instructions on how to proceed. After the operation has stopped, each of the valves on each pipeline system and at the manifold must be closed and the shore hoses or arms disconnected. In addition to taking the abovementioned measures, the crew must be informed of the situation and every effort made to prevent or eliminate further hazards.

If necessary, the Master may call the crew to security stations and order the ship's engine on standby. To ensure the crew and vessel are able to respond to any manner of emergency situations that could occur during cargo handling operations, the Master must instruct the Chief Officer to educate and train the crew in how to respond to emergency situations in accordance with the cargo operation manual. Such responses may include stopping the cargo pumps, closing valves and operating lifesaving and firefighting equipment, in addition to oil spill response equipment and materials. Furthermore, as part of the emergency response training, the Master – through the Chief Officer – is responsible for providing education and training in accordance with the vessel's contingency manual. Such education or training, when provided, must be recorded.

Precautions for cargo control

Preservation of the quality of cargo

In order to preserve the quality of the cargo carried onboard, proper cargo control is required. This may be achieved by maintaining secure segregation, heating and monitoring. Key to safeguarding the integrity of cargo is preventing the mixture of different grades of cargo. Accordingly:

1. Special attention should be paid to the prevention of cargo contamination whenever the vessel is loaded with two or more different grades of cargoes and when they are shifted between cargo tanks at sea; usually to adjust trim or heel or when an opening/closing test of valves is carried out, and
2. Attention should be drawn to cargo contamination, depending on the properties of cargoes, as a result of their vapours shifting via the vent lines.

Between different grades of cargoes with a large difference in Reid vapour pressure, a considerable amount of cargo may shift through the movement of vapours passing along the vent line.

Cargo heating

Careful temperature control should be exercised over cargoes liable to solidify with a high waxy content or viscous cargoes with a high pour point. Instructions on how to temperature control cargoes will always be provided by the shipper. These instructions must be followed diligently. Insufficient heating of cargo may cause delays in cargo discharge as well as blockages which may prevent cargo discharge or result in a large amount of cargo being left behind. With waxy cargoes, once the cargo temperature drops below the solidifying point, the cargo will solidify forming heavy layers at the tank sides and bottom. To remove, the tank must be heated to melt the wax. Only then can the cargo residue be safely pumped out of the tank. Alternatively, excessive heating of cargo may cause a deterioration in the quality of the cargo by vaporisation or separation of the lighter fractions of cargo. These vapours may then hang in suspension causing vapour locks or high-temperature trips of the cargo pumps during cargo discharge operations.

Measures to prevent vapour loss

Cargoes with a high vapour pressure that contains a large amount of volatile content may partially vaporise at sea, resulting in vapour losses. In order to minimise the amount of such loss, the following measures should be implemented:

1. **Inspection and maintenance of breather valves.** Breather valves should be regularly inspected and serviced, and their maintenance recorded. The normal set pressure should be maintained at all times to prevent the unnecessary release of vapours.
2. **Tank openings.** In order to maintain the air-tightness of tank openings, including access hatches, peepholes and tank cleaning holes, the entry points should be inspected and serviced at regular intervals to prevent leakage of vapours. Inspections and servicing must be carried out at least monthly during ballast passages or earlier if required. Again, these inspections and servicing intervals must be recorded.
3. **Water spraying over the deck.** When it is likely the tank pressure will exceed the set pressure of the breather valve, seawater may be sprayed over the deck plate to cool the upper deck and ship's side. This is done to lower the temperature and pressure of the ullage space.

Safety measures against cargo leaks

During a laden voyage, soundings should be taken of any compartments which adjoin the cargo tanks. This may include the pumproom, ballast tanks and cofferdams. The results of each inspection must be recorded daily. Internal inspections should be made, as necessary, to identify leaks at their earliest stage. Level gauges should be put in operation at least once every day while at sea, and ullage should be measured to check for unusual conditions. In the event it is noticed that cargo has leaked into the cofferdam, ballast tank, and/or other spaces, the crew must ensure all safety precautions are taken as per the physical and chemical properties of the cargo taking account of potential fire hazards, the toxic nature of the cargo, potential reactivity etc. Emergency measures must be taken to reduce/eliminate the risk of fire and explosion. If possible, remedial actions may be taken to control leaks and/or retransfer cargo back to the tank. Personnel involved in these operations must wear appropriate personal/respiratory protective equipment. Notification must also be made to vessel the owner and charterer, with a request for any specific contingency or advice. Care must be taken to ensure the hydrostatic stability of the vessel is not negatively affected.

Cargo release at sea and at the terminal

In the event cargo has leaked out to sea or at the terminal, the Master must be guided in principle by the company's contingency plan. If the physical properties of the cargo are designated as toxic, then immediate care and precaution must be taken to safeguard the health of all shipboard and onshore personnel. As an immediate response, the crew should aim to secure any inlets and dilute/reduce the toxicity of the cargo. The Master should take suitable measures to minimise the effect of cargo release into the marine environment/atmosphere. In all events involving accidental cargo release, the terminal and Port State authorities must be notified.

SAFETY MEASURES REGARDING METHANOL CARGOES

When handling a cargo of methanol, special precautions must be taken. Cargo tanks must be inerted/padded with inert gas, wherever fitted, with an inert gas generator. Where possible, the terminal should be contacted in advance to arrange for the connection of a vapour return line. If the weather conditions are cloudy and there is a potential for lightning at any point during the loading operation, it is worth considering whether the venting of vapours should be performed through the mast riser or a single vent, whichever can be secured the fastest. The terminal should follow suit. Venting of vapours through individual PV vents should be avoided as this takes longer to shut. During the transportation of methanol from one port to another, the pressure from the tanks should be released regularly to prevent the PV vents from lifting during lightning episodes.

> Under no circumstances is the release of methanol vapour to occur during periods of lightning.

OIL TANKER CARGO LOADING OPERATION PROCEDURES

The loading of oil cargoes into a tanker requires the utmost diligence in planning and the most careful consideration to ensure safe operation. The following principles are a summary of the procedures to be followed at the various stages of loading oil cargo.

Line up of the vent lines. Prior to the loading operation commencing, the cargo tank's inert gas inlet lines must be designated according to the assigned tank. This should be double-checked and confirmed at least twice before loading starts. The control of the key to the locking arrangements for the cargo tank inert gas inlet valve should be held in the person of the Chief Officer. For tanks which are required to be isolated by vapour (in accordance with the charterer's instructions), then individual inert gas pressures should be monitored at a minimum of every 4 hours.

Safety confirmations and clearance. Once the Chief Officer is satisfied that all preparations have been made in accordance with the cargo oil loading plan and the shore facility representative has confirmed that the facility is ready to load cargo, the Chief Officer may issue the order for the opening of the designated manifold valves. This signals the loading operation is to commence, always in accordance with the loading plan. Commence loading at a reduced rate (to avoid static generation), always monitoring the manifold back pressure at all times. The first loading tank should be documented in the "Tanker Cargo Logbook." The number of tanks being filled at the beginning of the procedure should be restricted to the absolute minimum. Ullage confirmation should be carried out to confirm the cargo oil is flowing as planned into the designated cargo tank. In the case of heated cargo, confirmation of the cargo temperature should be given to the Chief Officer as per the agreed value and within the charterer's instructions. Further, the loaded cargo temperature must always be within the vessel's design criteria (accounting for valve/tank coating limitations). Only once the Chief Officer has received reports that all contingent safety checks have been carried out and the correct status confirmed from all deck/pumproom watch stations to their complete satisfaction, may the Chief Officer issue the order to open the other loading tanks, whilst carefully increasing the load rate. It is critical at this stage that a close watch of the manifold back pressure is maintained, until completion of the settling down of the final maximum agreed loading rate. For safety, close communication must be kept between the ship and shoreside, until all parameters have stabilised. The loading cargo tanks' inert gas back pressure must be continuously monitored and adjusted to maintain a slight positive. Any unexpected fluctuations or changes in pressure must be recorded and the Chief Officer notified immediately.

Deck watch and personnel arrangement. The deck watch must check for oil leaks in the cargo area throughout the cargo oil loading operation. At the beginning of operations, tank and pumproom inspections must confirm there are no oil leaks from piping joints and that no oil is flowing into any of the ship's tanks other than the tank being loaded. It is critical that continuous monitoring of the loading tank oil level is maintained until the settling down of the shore flow rate. This includes monitoring all other (unused) tanks for any change in levels. After reaching the desired full loading rate and confirmation reports have been received from all stations at deck/pumproom watch (including the cargo piping and sea surface around the vessel) the Chief Officer may dismiss the off-duty crew and revert to the routine Watch Schedule. During loading operations, the deck watch must continue to monitor the manifold back pressure, especially when changing over valves/tanks, until ordered otherwise.

Leakage monitoring system. Cargo leaks, however small, must be rectified immediately. Leaks from the piping system, joints, and valves should be fixed and monitored. Any tanks not being loaded should be monitored to ensure there is no oil flow into the tanks, other than those designated as loading tanks. During loading operations, watch the oil loading pressure at all times, and monitor those sections where oil is likely to leak. Excessive vibration to the piping system must be investigated and rectified immediately.

Cargo loading rates. The vessel's maximum loading rate and maximum venting capacity must be posted in the cargo control room. This document should detail the rates for homogenous (i.e., for the entire vessel), group-by-group and tank-wise loadings. This information, based on calculations, should assist the Master in determining how fast the ship can safely load a particular cargo at a particular facility, taking account of the vessel's design parameters and the cargo involved. The Chief Officer should indicate, as per the loading plan, what rates are required at each stage of the operation.

Theoretical rates. The maximum flow rate into any single tank should be less than the maximum venting capacity of the tank in accordance with SOLAS. To allow for the generation of gas when loading, the venting rate may be taken as 125% of the oil loading rate. Maximum loading rates are affected by a number of factors, including the diameter of the manifold valve/line and the cross-section of the pipe. This factor can be calculated using the following formula:

> [m^2] × instant flow rate 7 [m/sec] × 3,600 [sec] = reference max. loading rate

Furthermore, the number of tanks being loaded at any one time, the capacity of the gas venting (main system) and the secondary gas venting capacity will also greatly influence loading and discharging rates.

Setting loading rates. The initial and maximum loading rates, topping-off rates, and normal stopping times should be calculated having regard to:

The nature of the cargo to be handled.

The arrangement and capacity of the ship's cargo lines and gas venting systems. The vent line pressure should not exceed that indicated by the manufacturer and must be closely monitored at terminals where loading rates are known to be high. The manufacturer's maximum vent pressure may be based on a rate for loading all tanks simultaneously; in which case, rates must be reduced accordingly when a smaller number of tanks are being loaded.

The ability and competence of the vessel's staff.

Integrity of the vessel and onshore loading/discharge system. The loading rate will be governed by the age, condition and reliability of the vessel's pipeline system and the gauging system.

Precautions to avoid accumulation of static electricity.

Any other flow control limitations.

Deballasting of segregated ballast

Before starting to de-ballast the segregated ballast tanks, always obtain the Berth (Loading) Master's permission first. In principle, de-ballasting operations should commence after the start of cargo operations. Deballast, as per the cargo plan, to achieve ample trim, especially towards the completion of de-ballasting operations. This eventuality should be planned well before the level in the cargo tanks near the topping-off ullages.

Recording of operations in the tanker cargo logbook

A number of items must be recorded in the tanker cargo logbook on an hourly basis. These include, but are not limited to, the loading quantity (rate). This is recorded to compare it with that of the terminal side. Regular ship/shore comparisons of loaded cargo figures should be carried out and any changes in difference investigated and reported. If the duty deck officer cannot account for the variation in rate, then the Chief Officer must be summoned immediately. Other data parameters which need to be collected are the manifold pressure/temperature, the ship's draft and trim, monitoring reports of levels in tanks not being discharged, and the stress and stability of the vessel and tank pressures. SBM/FSO position monitoring should also be carried out throughout the loading operation. The crew on watch must be briefed on the danger limits for the bearing and distance of the SBM/hawser.

Inerting

Oil tankers fill the void space above the oil cargo with inert gas to prevent the ignition of hydrocarbon vapours which can lead to explosion or fire. Oil vapours cannot burn in air with less than 11% oxygen content. The inert gas is often supplied by cooling and scrubbing the flue gas produced by the ship's boilers. Where diesel engines are used, the exhaust gas may not have a sufficiently low oxygen content in which case fuel-burning inert gas generators may be installed. One-way valves are installed in process piping to the tanker spaces to prevent volatile hydrocarbon vapours or mist from entering the other equipment. Inert gas systems have been required on oil tankers since the SOLAS regulations were enacted in 1974. The IMO publishes the technical standard IMO 860 which describes the requirements for inert gas systems. Other types of cargo such as bulk chemicals may also be carried within inerted tanks; however, the inert gas must be compatible with the chemicals carried.

Loading

When commencing the loading operation, it is imperative to start at a reduced rate to avoid static generation. Whilst simultaneously watching the manifold, it is important to monitor the back pressure at all times. Ullage confirmation should also be carried out to confirm the cargo oil is flowing as planned into the designated cargo tank. Only once all reports of safety checks have been confirmed from all deck/pumproom watch stations can the Chief Officer instruct the opening of other loading tanks and carefully increase the loading rate. A close watch of the manifold back pressure must be maintained until the maximum agreed loading rate has settled to a steady stream. Close communication must be maintained with the shore-side loading team until all parameters have stabilised. To prevent accidents, the loading cargo tanks' inter-gas back pressure may need to be adjusted to maintain a slight positive pressure at all times. This must be monitored for any fluctuations or abnormalities.

Unloading

When starting to ballast, the cargo pumps should be operated so that no oil is allowed to escape overboard when the sea suction valve is opened. Readers may wish to consult with the ICS/OCIMF publication *Prevention of Oil Spillages through Cargo Pumproom Sea Valves* for further guidance. With respect to the sequence of valve operations, the following procedures should be adopted when loading ballast into non-inerted tanks which contain hydrocarbon vapour. To begin with, the tank valves should be the first valves opened. The initial flow of ballast should be restricted until the longitudinal are covered or, if there are no longitudinal until the depth of the ballast in the tank is at least 1.5 metres (4.9 feet) in depth. These precautions are required to avoid a geyser effect which may lead to the buildup of an electrostatic charge in a mist or spray cloud near the point where the ballast enters the tank. When a sufficient charge exists the possibility of a discharge and ignition cannot be excluded. The Chief Officer should also prepare a watch schedule and Person in-Charge (PIC) list for oil transfer and discharge operations. Prior to the commencement of the discharge operation, the Chief Officer should conduct a "Pretransfer cargo safety meeting" with all personnel involved in the pending unload operation.

Tank cleaning

Tanks must be cleaned from time to time for a number of reasons. One key event is when the product to be carried changes. Also, when tanks are to be inspected or maintenance must be performed within a tank, it must not only be cleaned but made gas-free. On most crude oil tankers, a special crude oil washing (COW) (Figure 6.2) system forms part of the cleaning process. The COW system circulates part of the cargo through the

Figure 6.2 Tank cleaning.

fixed tank-cleaning system to remove wax and asphaltic deposits. Tanks that carry less viscous cargoes are washed with water. Fixed and portable automated tank cleaning machines, which clean tanks with high-pressure water jets, are widely used. Some systems use rotating high-pressure water jets to spray hot water on all internal surfaces of the tank. As the spraying takes place, the contaminated liquid is pumped out of the tank. Once the tank is cleaned, provided that it is going to be prepared for entry, it will be purged. Purging is accomplished by pumping inert gas into the tank until all hydrocarbons have been sufficiently expelled. Next, the tank is gas freed by blowing fresh air into the space with portable air-powered or water-powered air blowers. "Gas freeing" brings the oxygen content of the tank up to the required volume of 20.8%. The inert gas buffer between the fuel and oxygen atmospheres ensures they are incapable of ignition. Specially trained personnel monitor the tank's atmosphere, using hand-held gas indicators which measure the percentage of hydrocarbons in the air. After the tank is gas-free, it may be further hand-cleaned using a manual process known as "mucking." Mucking requires expansive protocols for entry into confined spaces, protective clothing, designated safety observers and potentially the use of airline respirators.

Purging and gas-freeing

Gas-freeing for cargo tank entry

Cargo tank entry must not be permitted unless the oxygen content is 20.8%–21% and the hydrocarbon vapour content is less than 1% of the lower flammable limit (LFL). When effecting entry into the cargo tank, or indeed any enclosed space, it is imperative the company's "Procedure for Entry into Enclosed Spaces" is followed without exception. This also means applying for having the required permits to work (PTW) authorised and signed by the Master and/or Chief Officer. If the previous cargo contained hydrogen sulphide (H_2S) or any other toxic contaminants which could evolve toxic gases (e.g., benzene, toluene and Mercaptans), the tank should be checked for the presence of such gases. As part of the necessary preparations, it is always good practice to refer to the company's Standard Operating Procedure (SOP) relating to protecting against the effects of toxic gas hazards and carrying out "hot work" inside tanks within the "dangerous area" (wherever applicable).

Gas-freeing or purging for the reception of cargo

If the intention of gas-freeing or purging operations is to protect the next cargo to be loaded from contamination from the previous cargo, usually oil hydrocarbon gas, always refer to and use the gas content as indicated by the charterer.

Purging and gas freeing onboard tankers

Inert gas purging

If inert gas purging is being carried out by the displacement method any dense concentrated hydrocarbon layer at the bottom of the tank is expelled in the early stages, followed by the remainder of the tank atmosphere as it is pressed downwards by the inert gas. If there is a uniformly high concentration throughout the tank, for example, after product washing, the product concentration of the vented gas will remain high throughout the purging process until the inert gas reaches the bottom of the tank. If inert gas purging is being carried out by the dilution method the gas concentration at the outlet will be highest at the start of the operation, before falling as the cleaning procedure continues.

Gas freeing

It is recognised that gas freeing is one of the most hazardous activities in tanker operations. This is true whether gas freeing for entry, for conducting hot work, and/or for cargo quality control. The cargo vapours that are to be displaced during gas freeing are highly flammable; therefore, good planning and firm overall control are essential. The additional risk from the toxic effect of cargo vapours during this period cannot be overemphasised and must be impressed on every member of the cleaning crew. It is therefore essential that the greatest possible care is exercised at all times when conducting gas freeing activities. It is strongly recommended that gas freeing is avoided as much as possible in order to reduce the risk of injury.

Gas freeing for entry without breathing apparatus

In order for the tank to be gas free for entry without needing breathing apparatus, the tank or space must be thoroughly ventilated until tests confirm that the cargo vapour concentration throughout the compartment is less than 1% of the LEL. The oxygen content must also be a minimum of 20.8%–21% by volume. Furthermore, there must be no confirmed presence of H_2S, benzene or any other toxic gases present. Before entering a tank without breathing apparatus, the atmosphere in the tank must be checked and confirmed as safe for entry by a suitably qualified and experienced person (SQEP).

Procedures and precautions for gas freeing

The following recommendations should be applied with respect to gas freeing operations. All tank cleaning operations must be adequately supervised by a SQEP member of the ship's crew. Before gas freeing activities commence, all personnel onboard the ship should be notified that gas freeing is about to begin. "No Smoking" regulations should be implemented and enforced.

Signage should be placed in conspicuous locations around the vessel. The instruments to be used for gas measurement should be calibrated and tested in accordance with the manufacturer's instructions before starting the gas freeing operation. Any faulty equipment must be removed and quarantined as per the ship's SOP. Replacement equipment must be checked and tested before being brought into service. Sampling lines should, in all respects, be suitable for use with, and impervious to, the gases present in the tank. All tank openings should be kept closed until the actual ventilation of the individual compartment is about to commence. Venting of flammable gas(es) must be carried out in accordance with the vessel's approved procedures.

Where gas freeing involves the escape of gas at deck level or through hatch openings, the degree of ventilation and number of openings should be controlled to produce an exit velocity sufficient to carry the gas clear of the deck. Intakes of central air conditioning or mechanical ventilation systems should be adjusted, if possible, to prevent the entry of gas or vapours, by recirculating air within the spaces. If at any time it is suspected that gas is being drawn into the accommodation, the central air conditioning and mechanical ventilation systems must be stopped immediately and the intakes covered or closed. Window-type air conditioning units which are not certified as safe for use in the presence of flammable gas, or which draw in air from outside the vessel's superstructure, must be electrically disconnected and any external vents or intakes closed. Gas vent riser drains should be cleared of water, rust and sediment, and any steam smothering connections tested and confirmed as satisfactory. If several tanks are connected by a common venting system, each tank should be isolated to prevent the transfer of gas to or from other tanks. Where cargo vapours are found to persist on deck in high concentrations, then all gas freeing should be stopped. Furthermore, if wind conditions cause funnel sparks to fall on the deck, gas freeing should cease with immediate effect.

Tank openings within enclosed or partially enclosed spaces, such as under the forecastle, should not be opened until the compartment has been sufficiently ventilated by means of openings in the tank that are outside those spaces. When the gas level within the tank has fallen to a minimum of 25% of the LEL or less, openings in enclosed or partially enclosed spaces may be opened to complete the ventilation. Such enclosed or partially enclosed spaces should also be tested for the presence of gas during this subsequent ventilation.

Chapter 7

Inspection and maintenance of cargo handling equipment at sea

GENERAL

The cargo remote control system should be inspected and serviced in accordance with the manufacturer's operations manual. This includes the electric system, including voltage, adjustment of contact points, grounding, testing of indicator lamps and operation of solenoid valves; the hydraulic system, including hydraulic pressure, hydraulic oil sump level, accumulator gas pressure, fouled hydraulic oil, contamination with water, sampling analysis, fouled strainer, activating pressure of the pressure relief valve, leak for the hydraulic pipelines and pressure cycle; and the air system, such as air pressure, presence of leaking points and presence of drains. In addition, if each equipment has an alarm system, its functions should be checked.

INSPECTION AND MAINTENANCE PROCEDURES

Maintenance of cargo valves

By paying attention to the operation of valves related to cargo operation systems and their leaks to prevent accidents, such as cargo contamination and overflows, the Chief Officer must provide maintenance service to keep them in good working condition. Before the commencement of cargo handling operations or on other occasions, the Chief Officer should conduct an operational test on valves, if necessary, to check for unusual conditions. The results must be recorded. Hydraulically driven valves require special attention in terms of the operational speed in order to prevent a pressure surge.

Maintenance of cargo pipelines

Cargo pipelines should be visually inspected and checked for leaks by applying positive or negative pressure at regular intervals. Any indication of leaks must be investigated in a timely manner with repairs effected without

DOI: 10.1201/9781003505044-7

undue delay. The test pressure applied to pipelines should be at least 1.5 times the maximum allowable working pressure (MAWP).

Cargo tank bulkheads

Tank bulkheads should be checked for indications of stress and leaks at every opportunity. Any indication of leaks must be investigated in a timely manner with repairs effected without undue delay.

Cargo pumps

All safety devices and systems associated with the ship's cargo pumps must be regularly tested. This includes confirming the emergency stop system of the cargo pump is functional at all times. The relief valve of the stripping pump should be pressure-tested when the vessel is in drydock or at any other convenient occasion.

Cargo vent system

The flame arrester fitted at the end of the vent line should be removed at least once every three months for cleaning. When inspection and maintenance services are provided, these should be recorded. The P/V valve of the vent line, cargo tanks and the P/V breaker should be regularly inspected and maintained as per the ship's planned maintenance system (PMS). Each P/V valve is to be numbered and a record kept of all maintenance for each valve maintained in the PMS. The correct maintenance of these valves is essential to the safe operation of the vessel. Each individual P/V valve on the cargo tanks is to be inspected visually for the condition of flame screens, any leaks and free movement. Each P/V valve is to be dismantled, overhauled, flame screened on the vacuum side and renewed, and the P/V valve tested on the test bench over a six-month cycle. This is to be done on the ballast voyage with the tank open to the atmosphere and all supply valves to the tank shut. On reassembly, valve tightness is to be tested using soapy water. Flame arrestor gauzes fitted on the P/V valves are to be checked prior to cargo operations to ensure that they are free from damage and the presence of polymerised substances, which may prevent freedom of vapour flow.

Cargo heating pipelines

The cargo heating pipelines should be checked prior to cargo loading and, where necessary, serviced in accordance with the PMS and manufacturer's instructions. This is to ensure all heating pipes are kept in good working order. Whenever a section of pipe is repaired, the cargo heating pipe should be cleaned by blowing steam through the pipe until no further cargo is discharged. If the repair of the cargo heating pipe is beyond the scope of the crew's capability, the heating valve and drain valve of the pipe should be closed and quarantined against use.

Lighting and electric systems for upper decks and pump rooms

The lighting system and electric wiring should be regularly tested and checked for lamp failures and electric leaks. Special attention must be paid to the condition of the glass casing for lighting and sheaths of cabling leading to masts and posts. If any damage is found, these should be repaired immediately.

Cargo samples, storage and disposal

Cargo samples which have to be kept onboard must be stowed in a designated space situated in the cargo area or, exceptionally, in another authorised area subject to the approval of the vessel's Flag State administration. The storage space must be cell divided in order to avoid the shifting of sample bottles out at sea and made of material fully resistant to the different liquids stowed. The space must be equipped with adequate ventilation arrangements. Samples which react with each other must not be stowed in proximity to each other. Depending on the prevailing circumstances, the samples should be retained for at least one year, by which time notification of any claim should have been made. That said, it is recommended that samples are retained for more than one year if there is reason to believe a claim may be lodged at a later date or if there is any doubt concerning the product quality during the voyage. In most cases, cargoes will be loaded and delivered without incident, and, in these cases, the storage and disposal of samples is a matter of following company procedure.

HANDLING OF WATER BALLAST

Taking on water ballast

Cargo tanks may be ballasted in any of the following cases:

1. When it is necessary to do so in adverse weather to secure the safety of the vessel,
2. When it is necessary to do so to pass safely under a bridge or other suspended obstruction, and/or
3. When local port or canal regulations require a specific draft to ensure the safety of the vessel.

Selection of tanks to fill with water ballast

A number of factors must be considered when selecting which tanks to fill with water ballast. Tanks to be filled with water ballast should be selected in such a manner that the safe navigation of the vessel is ensured, and the restrictions outlined in points (1), (2) and (3) above are complied with. For the purpose of carrying additional ballast in severe weather conditions, ballast tanks may be selected freely within the limit of restrictions on ballasting as provided above.

Precautions for ballasting

Every precaution should be taken when filling dirty ballast tanks with seawater. The overarching concern is to prevent pollution by discharging contaminated ballast water at sea. When ballast water is to be discharged to an onshore reception facility, this must be done in accordance with the latest instructions. Whenever a cargo tank is filled with water ballast, the Chief Officer must record the event in the Oil Record Book on every occasion.

Solidification in the cargo tanks can occur when solidifying cargoes are stowed adjacent to "cold cargoes" or cold ballast water in adjacent spaces. Cargo tank bottoms must therefore always be checked for hard factions, at regular intervals throughout the voyage and always prior to arrival in the port of discharge. Consideration has to be given to minimise the temperature reduction of the heated cargo whilst taking ballast into the adjacent tanks. Ballast intake should be planned accordingly in order to avoid solidification whilst carrying heated cargo. Any surface area that is exposed between the cargo and ballast must be minimised. If possible, there should be no exposure of the ballast and cargo via the ship's structure. This is to avoid any cooling effect which ballast water may have on the cargo, resulting in the solidifying of the cargo. Any remaining ballast is to be removed only after the cargo has been fully discharged from the concerned cargo tank. When ballasting double-bottom tanks or side tanks with a double bottom, the level of the ballast should be below the deck head (i.e., the cargo tank bottom) until the cargo is fully discharged. When carrying cargoes liable to solidification (such as palm oil, beef tallow, fatty alcohol and phenol) double-bottom tanks must be filled with ballast water in a controlled manner and never filled to its full limit. Ideally, the quantity of ballast should be adjusted to less than 70% of the ballast tank capacity.

Discharge of water ballast

Restrictions on deballasting

Any discharge of water ballast must be carried out in a controlled manner which complies with the MARPOL regulations. In summary, the following restrictions usually apply to all vessels including those engaged in the carriage of oil and chemical products:

1. **Discharge of water ballast from segregated ballast tank.** Water ballast in segregated ballast tanks should be discharged only after it is ensured that no oil/cargo is floating on the water's surface,
2. **Discharge of water ballast from cargo tank.** The discharge of clean or dirty water ballast from cargo tanks must be executed in accordance with the respective procedures as follows:
   a. **Discharge of clean ballast.** The discharge of clean ballast should be made following the same procedure as that for de-ballasting from segregated ballast tanks mentioned above. It is mandatory to use the oil discharge monitoring equipment (ODME) when discharging ballast from the cargo tanks in compliance with MARPOL 73/78 for Annex I cargoes and for oil-like substances.
   b. **Discharge of dirty ballast.** In case the last cargo carried was as per Annex I of MARPOL 73/78. The discharge of dirty ballast should be made according to the following conditions unless it is transferred to a shore reception facility:
      i. The tanker is proceeding en route,
      ii. The tanker is not within a special area (Mediterranean Sea, Baltic Sea, Black Sea. Gulfs, Red Sea, Gulf of Aden, Antarctic Ocean),
      iii. The vessel is more than 50 nautical miles from the nearest land.
      iv. The instantaneous rate of discharge of oil content does not exceed 30 litres per nautical mile,
      v. The total quantity of oil discharged into the sea does not exceed 1/30,000 of the total quantity of the particular cargo of which the residue formed a part,
      vi. The tanker has in operation, an oil discharge monitoring and control system and a slop arrangement, and
      vii. Discharge is made from above the waterline.

      Ballast in cargo tanks other than slop tanks may be discharged into the sea from below the waterline after it is ensured that the position of the oil/ water interface is checked before ballast discharge and the oil layer above it may not find its way into the sea when it is discharged by gravity.
3. **Discharge of water ballast from slop tanks.** When separated water is discharged for decanting from a slop tank; such discharge should be made from a position above the waterline without fail. In other cases, the discharge should be made in the same way as the discharge of dirty ballast mentioned above.

Deballasting when special regulations of the terminal of the port of call are in force

When deballasting is restricted by special regulations at the terminal or port of call, such regulations, in addition to the restrictions provided above in points (1) and (2), must be complied with. When water ballast is discharged, the restrictions on deballasting should be observed without deviation. When clean ballast or segregated ballast is discharged, in particular, special care should be taken to avoid inadvertent pollution of the sea surface with strict watch kept on the waters immediately surrounding the vessel. Attention should be drawn to the fact that even when segregated ballast is discharged, oil/cargo may mix into the ballast water through holes in the ballast line running through a cargo tank or bulkhead. To ensure compliance, the Chief Officer must record every instance where the vessel deballasts from the cargo tanks. Moreover, any decanting of the slop tank must also be annotated in the ship's Oil Record Book. Every instance where the vessel discharges cargo tank cleaning water must also be recorded in the Cargo Record Book.

Inspection and maintenance of equipment related to ballasting/deballasting

The following matters should be borne in mind when carrying out inspections and maintenance of equipment related to ballasting/deballasting equipment:

1. **Equipment also used for cargo handling.** All ballasting/deballasting equipment is also used for cargo handling,
2. **Oil discharging monitoring and control system.** The oil discharge monitoring and control system in the discharge of water ballast is a statutory device and should be inspected and maintained. It should be operated in accordance with the oil discharge monitoring and Control System Operational Manual. The oil discharge monitoring and control system should be approved for the use of oil-like substances and various other chemicals that the ship is certified to carry in accordance with Annex I of MARPOL 73/78, and
3. **Oil/water interface detector.** The oil/water interface detector, which is a statutory device, should be regularly inspected and maintained. It should be kept in the strict custody of the Chief Officer at all times.

Action when the oil discharge monitoring and control system is malfunctioning

When the oil discharge monitoring and control system malfunctions, the disposal of ballast water and/or decanting must be stopped immediately. The cause of the malfunction should be investigated and any remedial actions performed without undue delay. Where the oil discharge monitoring and control system cannot be fixed by the ship's engineers, this must be reported to the onshore vessel manager who will arrange for a suitably qualified and experienced person (SQEP) technician to attend the vessel at the earliest possible opportunity. The incident must be recorded accurately in the vessel's Oil Record Book.

Chapter 8

Cargo measurement and heating systems

GENERAL

The accuracy required of oil tanker and chemical carrier level gauges is inevitably high because of the nature and value of the cargoes carried by these types of vessels. To limit personnel exposure to chemicals and their vapours during loading, unloading and carriage at sea, the IBC Code specifies three methods of gauging the level of a liquid in a tank: open, restricted or closed. The method stipulated is in accordance with the health hazard of the product. Many chemical cargoes may not be gauged by manual dipping because to do so requires an opening to the atmosphere during operation. This means the use of completely closed gauging systems is necessary to prevent the egress of gases or vapours. Examples of closed systems include float gauges or radar systems. Indirect measuring methods such as flow metering may also be used. All toxic cargoes require either restricted gauging or closed gauging. Many more chemicals, although still hazardous, do not require quite such rigorous controls, and restricted gauging accepts that an exceedingly small amount of vapour may escape during gauging. One such example is using a sounding pipe which reaches far down into the tank. Other cargoes can be gauged through openings in the ullage space. This is called "open gauging."

CARGO MEASUREMENT

Ullage

"Ullage" is the measurement of space between the surface of the liquid in a tank and the top of the tank's inner surface. Oil and chemical tankers usually carry a document called the ship's *ullage tables*. The ship's ullage tables illustrate the internal volume of each tank as measured from some reference point, for example, the lip of an ullage hole. In oil tankers, ullage is left in order to leave room for expansion when the oil is heated to a higher

DOI: 10.1201/9781003505044-8

temperature before discharge. Oil can also expand with changes in atmospheric temperature therefore it is customary for oil tanks to be loaded to a maximum of 98% capacity. The reverse of ullage is "sounding." This is the depth of liquid in a tank measured from the liquid surface to some reference point at the bottom of the tank. Like ullage, these measurements are recorded in documents called *sounding tables*. The expression "sounding" covers the free space left in the tanks after loading liquids in bulk. Some ships carry both *sounding tables* and ullage tables, or one or the other, depending on the shipowner's preferences.

Float gauges

Float gauges are a type of closed gauge and consist of a float which rises vertically as the volume of liquid in the tank increases. It is attached by a tape to an indicating device which enables local reading. In most cases, there is added provision for a drive mechanism which facilitates remote readouts. With respect to float gauges, there are several points which need to be considered and/or borne in mind, especially when the vessel is underway. Primarily, the floats should be secured when the vessel is at sea, except of course when measurements of the tank contents are being taken. If the float is left unsecured at sea, it will certainly sustain damage due to the free movement of the cargo. This will inevitably lead to inaccurate readings. The second point to keep in mind is that remote and local readings should always be compared to identify discrepancies. In the event discrepancies are identified, these should be investigated immediately. In the event there is a material (i.e., actual) discrepancy, the Master must be informed without delay. Third, readings may need to be corrected to allow for tape and tank expansion or contractions, as well as changes in ship trim and heel. Trim and heel tables are normally provided for this purpose. Fourth, tapes should be checked regularly to confirm the free vertical movement of the float. If the float is damaged, it should be removed from service, isolated in quarantine, and then disposed of onshore in accordance with the ship's standard operating procedure (SOP). Needless to say, the float must be replaced with a non-damaged float. Particular care is also necessary when handling the rewind mechanisms as these are carefully balanced; if obstructed, the gauge readings will be inaccurate giving false measurements. The fifth point to note concerns when tapes need to be renewed or a gauge reassembled after maintenance. In this situation, the manufacturer's instructions must be consulted. Furthermore, allowance should be made for the level at which the float begins to lift. The sixth and final point is that parts should be securely assembled at all times. Special care is necessary with tape-to-float and tape-to-reel attachments.

Procedures for measuring (closed and restricted)

When vessels are fitted with vapour control valves (VCVs), portable electronic gauging equipment can be used to measure free water, petroleum liquid levels and temperatures. They may also be used for measuring the remaining onboard (ROB) or onboard quantity (OBQ) of liquids. Special portable measurement units (PMU) and/or techniques may be used for taking samples and for measuring non-liquid ROB/OBQ. Use of this equipment requires careful observance of the safety procedures outlined in the International Safety Guide for Oil Tankers and Terminals (ISGOTT), International Maritime Organisation (IMO), Inert Gas Systems (IGSs) and other applicable International Chamber of Shipping (ICS), Oil Companies International Marine Forum (OCIMF) publications and manufacturer's instructions. Prior to boarding a vessel equipped with VCVs, always determine the manufacturer and size of the VCVs so that compatible equipment or adapters can be brought for use onboard. In addition, before gauging, always verify that all cargo operations have been stopped, and no cargo is being transferred. The IGS pressure in the cargo tanks must be lowered to sufficiently to minimise the potential for vapour loss. The gauging equipment must be calibrated, and the calibration/verification log reviewed. The equipment must be checked and confirmed as free of breaks, kinks and signs of wear which might affect the measurement accuracy. The equipment must be clean for the product to be measured, i.e., the numbers and graduations on the tape are legible; and where appropriate, batteries are charged and replacements available. It may be noted that for best accuracy, trim and list should be eliminated. When both conditions exist, every effort should be made to eliminate at least one condition, preferably a list. Conditions of trim and list must be noted, and corrections made for their effect on measurements/volumes.

CARGO HEATING

All chemicals carried at sea are heated by hot water and not steam. The tank heating coils which are located at the exit point of the tank must also contain hot water and not steam. The temperature of the water in the coils is adjusted by the return valve from the tank. The steam inlet valve to the tank must be 100% open at all times. Heating is carried out for a number of reasons. This includes reducing the viscosity of the oil or chemical products. Viscosity is measured in centistokes (cst). Water is defined as one cst. As the temperature of the product rises, the viscosity of the product reduces at the average rate of 2% per degree in temperature. Heating is also done to reduce the pour point of the product. The pour point is 3°C (37.4°F) over the temperature where the product begins to coagulate. Similarly, heating the product helps to reduce the cloud point. This is the temperature where crystalised solids begin to form in suspension inside the liquid. If left suspended in the fluid, the crystalised solids will settle on the tank's bulkheads. Some cargoes, such as caustic soda, require heating to avoid crystallisation from occurring. Alternatively, a number of cargoes such as cyclohexane require heat to prevent freezing.

Heating is also widely used to support normal cargo loading and discharging operations. By heating the cargo, this can avoid pre-washing the tanks at the port of discharge. Moreover, heating may be required to comply with the MARPOL regulations for prewashing; for example, phenol tanks must be prewashed where the discharge berth temperature is below the melting point (MP) plus 10°C (MP + 50°F). Heating helps to increase water solubility; phenol, for instance, is soluble in water at a temperature of 60°C (140°F). Related to the point above regarding the presence of crystalised products, heating the product helps to reduce the presence of unpumpable elements at the tank sump and reduce "clingage" (which excludes coagulation at the tank bottom). Heating the product reduces the delivery pressure of the ship's centrifugal deep well pumps, which leads to quicker and more efficient discharge and loading. Last of all, heating the product precludes claims from the ship's charterers who want a particular voyage and discharge heat. Underheating and overheating the product during transit will both result in claims.

For tank heating coil systems, the size of the main return line is usually smaller than that of the steam inlet line. This is because the return line is supposed to contain only water whereas the main inlet line is supposed to contain steam. For steam winches, the opposite is true as the return line is bigger to compensate for the expanded steam. From the engine room boiler, steam is pumped on deck through the main inlet steam pipe. Before it enters the deck, it has to pass through a pressure control valve or PCV, with a 1 bar to 7 bar range. For safety, there will be a small bypass line around this valve for initial warming prior to starting. On smaller sized chemical tankers, this bypass must be operated for at least 45 minutes before the PCV is opened. Heating must be started slowly to reduce the impact of thermal stresses. When not in use the steam coils should be full of fresh water unless the cargo is water-reactive, in which case it must be blown dry and blanked. For initial starting, the PCV is put at 1 kilogram and then slowly increased. Afterwards, all drain cocks in the steam inlet line are opened until the water is ejected and steam flows out.

Only water should be returned to the engine room through the return line; this is primarily an energy conservation measure to save money. This water must be pure and should not contain any cargo or contaminants. There is an inspection chamber (called a *siphoning drum*) where the initial one hour of return water must be monitored. Some chemical tankers are fitted with a glass port which can be used to check for floating-insoluble cargoes, to check the odour from the vent to check for volatile soluble cargoes and to inspect the drain at the bottom of the tank to check for sediments or high specific gravity (SG) type insoluble chemicals. When carrying heated cargoes, the return siphoning drum content must be inspected daily to check for any traces of cargo residue. If any cargo residues are found, then the heating coils must be drained individually to establish the cause. This coil can then be blanked off. The drain valves and siphoning drum must be checked twice daily for any signs of ingress of cargo. The heating of the ship's cargo tanks should be regulated also by the number of active coils from the inlet manifold. The temperature of the return line has to be taken regularly to maintain a constant temperature and to avoid overheating or underheating. It must be noted that centre tanks with a double skin are usually heated to a lower temperature than the wing tanks which have a cold interface with ballast. Critical to the integrity of the heating system is knowing when and how to stop the heating, especially to avoid trapping high-pressure steam in the lines, which on cooling will cause a high vacuum to build up. This can lead to corrosive, explosive and/or toxic chemicals to be sucked into the coils.

Depending on the specific design and functional operation of the system onboard, the general procedure to be followed is set out here. Begin by closing the inlet valve, then open the drain cock in front of the return valve. Close the return valve. Given only one single leaking coil can contaminate the other clean coils via the return manifold, steam blow contents must always be checked for pH. If the steam heating coils fail – which may or may not result in the declaration of an emergency – use adjacent heat including live steam in the ballast water around the tank. It is worth noting that when no heating is carried out, the temperature inside the tanks will usually drop by 1–2°C (33.8–35.6°F) per 24-hour period. When heating, start by raising the temperature by between 5 and 6°C (41–42.8°F) per 24-hour period; do not exceed 6°C (42.8°F). Bear in mind that many cargoes perish due to indiscriminate overheating. Temperatures that are too high can result in an unacceptable vapour pressure (VP) or chemical/physical change, which can be permanent. When vegetable and/or animal oils are overheated carbon monoxide is formed. This gas is highly toxic and can adversely affect the sweeping party. Overheating lubricating oils can cause oxidation to occur in the cargo resulting in a change of colour whereas overheating molasses can cause thermal decomposition and total destruction of the product. The next step is to ensure the cargo is at the discharge heat four days prior to arrival and discharge.

To expedite the process, it may be possible to remove the cold ballast interface. The heating coils of tanks not required to be heated must be blanked on both entrance and exit and a logbook entry made to that effect. In order to avoid confusion, heating coils should only be used when the cargo requires heating. Any other cargoes onboard should have their coils blanked on both the inlet and return side. This is also of paramount importance if the cargo is inhibited, subject to polymerisation (styrene monomer) or has a violent reaction with water (sulphuric acid) etc. Tanks which require heating must be pressure tested prior to loading with a logbook entry made to that effect. In the event any of the coils leaks, immediately stop discharging/loading operations and inform the ship's Chief Engineer and senior deck officers. Commencement of discharging/loading may only begin once the Chief Engineer has effected permanent repairs using stainless steel welding. On a final note, if a heated valve is found to be too tight, allow it to cool. Never force the valve as this may result in damage. Some ships have deck-mounted heat exchangers, such as the Framo pump, which discharges into a heater at about 60 bars. The two methods for testing the steam coils onboard the ship are hydraulic testing using water and steam pressure testing. It is possible to carry out pneumatic testing using air however this method is not considered dependable and should therefore be avoided wherever possible.

Hydraulic testing using water

Hydraulic testing of heating coils is time-consuming and cannot be carried out in all tanks on a regular basis. Even so, the hydraulic testing of heating coils must be carried out the by ship's engineers at a minimum once a year or as dictated by company policy. It is recommended that tank tests are staggered so they are not all due at the same time. Hydraulic testing must also be carried out by the Chief Engineer after any permanent repairs have been effected to the heating coils. The results of all hydraulic tests must be recorded in the logbook and vessel's the planned maintenance system (PMS). In accordance with good practice, steam testing of the coils should be carried out before the loading of each heated cargo. To ensure efficient operation, thermal oil heating coils should be hydraulic tested, using oil, every two years. If defects occur, then the coils must be tested annually. The coils must also be tested following any repairs. All coils must be hydraulically tested when they have been replaced following disassembly for coating work etc. In a docking situation, this should be carried out prior to departing the dockyard. If this is not possible, then testing should be scheduled for as soon as it is practical. Any cargo damage caused by steam coil leaks or the ingress of cargo into the steam coils must be prevented. Cold hydraulic testing can be performed quite effectively using a Graco pump and a 200-litre fresh-water drum to pressurise the system to 12 kilograms. When carrying out inspections, it is important to check the riser piping for the presence of loose nuts from the steam coil clamps. Testing must be done coil by coil.

It is worth noting hydraulic testing is the more severe of the two test methods due to the fact that when using steam, heat expansion can cause small leaks to seals and also results in higher pressures (1.5 times) than the working pressure used. When carrying out hydraulic testing, it is important that the tank is completely dry, and the cold-water ballast interface at the tank bottom is removed. This will make the detection of exceedingly small leaks much easier, i.e., the presence of water at the bottom of the tank is an indication of a leak. To perform the test, install blanks to provide the required test boundary. Connect the hydraulic test pump to the coils and install a suitable pressure gauge. Fill the coils with clean fresh water. Build up the pressure in the coils to 1.5 times their maximum rated working pressure. Close the hydraulic pump valve to prevent a drop in pressure. Monitor the pressure for at least 30 minutes, without providing additional pumping. If a leak is indicated by a pressure drop, then the entire system must be checked until the leak is found. After 30 minutes, thoroughly examine each coil individually. The thorough examination must consist of a close visual inspection as well as a tactile (touch) test by feeling by hand under the coils in order to detect any pinhole leaks. Such pinhole leaks may not be indicated by a pressure drop on the manometer and can also be extremely hard to detect by the naked eye. If any leaks or defects are identified, these must be repaired and properly documented before the system can be authorised for use. Repeat the test procedure following each repair.

Steam pressure testing

The second method for testing the coils is to use steam pressure from the ship's boiler. Misleading results may be obtained from the steam test due to steam condensation in the coils if the test is not conducted properly. In all cases, the maximum boiler pressure must be used. The coils to be tested should be exhausted to the atmosphere until live steam is issued from the return. Close the return against the steam. The steam pressure must be held for 30 minutes prior to effecting an inspection. Alternatively, another two-in-one (DUAL) method may be used where the return valve is closed. The steam inlet is then opened at 7 kilograms. This will then provide an easy hydraulic test on the coil due to the trapped condensate in the line. Inspect the coil for any indications of water leaks. Following the above, the return valve is opened, and steam is allowed to flow through the coil via the steam trap. Once 30 minutes have elapsed, the coils are to be inspected for leaks. This can be achieved by shining a torch light along the coil; any escaping steam will indicate a fissure. If a leak is detected and a section of piping has to be repaired, a proper hydraulic test should be carried out once the repair is completed. Install blanks to provide the required test boundary, i.e., isolate only the coil to be tested. When carrying out visual inspections for leaks, it is important to look closely for signs of corrosion. Corrosion leaks may require further close examination of the coils to determine the amount of damage. Particularly bad corrosion may mean the whole section has to be cropped and renewed. All corrosion problems must be reported to the chemical operator immediately.

Heating coils not in use must be steam pressure tested every three months and the results logged and recorded in accordance with the vessel's SOP. If heating is not to be done the coils must be blanked, with a logbook entry made to record that fact. In the event the Master is unable to comply with the vessel's charterers' heating instructions, they must inform the chemical operator well in advance. Always ensure that there is sufficient bunker onboard to run the boiler. Under no circumstances should cargo exceeding tank lining resistance table temperatures be accepted for loading. Furthermore, always be aware of the maximum cargo temperature stipulated by the charter party. It is the responsibility and duty of the Chief Officer to record the temperature twice daily; the critical heating log must also be signed by the Chief Officer daily; the sign-off signature must never be postponed to the next day. A signed copy should be hung on the cargo control room (CCR) bulkhead for the Chief Engineer to monitor. Depending on the ship's SOP, the steam valves of wing tanks and forward tanks may be opened more regularly. The percentage (%) of the return valves opened must be entered in the log column.

The Chief Engineer must log down each day the bunkers used by the boiler solely for heating the cargo. Heated cargo temperatures must be taken at three levels from inside the tank. As the cargo is discharged, the steam heating into the tank must be reduced. Heated cargoes must be stripped immediately – they **MUST NEVER BE POSTPONED**. Furthermore, an action plan should be drawn up as part of the pre-load/pre-discharge meeting, to melt solidifying cargoes from inside the pipelines and to prevent valves from freezing in place. If steam hoses have to be used the crew must use PPE and heat-resistant gloves. In this instance, steam must be applied to the underside of the pipes. It is usually more effective to use hot water on burlap-lagged pipes, especially at bends. If the pump impeller is frozen, steaming into empty DB via the sounding pipe can be considered but always inform the chemical operator first. Before discharging a high MP cargo, it is worth considering recirculating the cargo before opening the tank delivery and manifold valve.

Thermal oil heating on chemical tankers

Thermal oils are heat transfer fluids that transfer the heat from one hot source to another process. This could be from a combustion chamber or from any exothermic process. The main application is in fluid-phase heat transfer. They are available in chemically different forms, such as synthetic oils, which are aromatic compounds; petroleum-based oils, which include paraffin; and synthetic glycol-based fluids. Thermal oils are available in a wide range of specifications to suit the needs of various processes. Currently available thermal oils have a maximum upper temperature limit of around 400°C (752°F) however there are thermal oils that are used in cryogenic processes which can withstand temperatures of –100°C (–148°F). In terms of its application onboard tankers, a heat transfer medium may be required where the direct heating of a process is not possible. Historically, steam was always the prime choice as a heat transfer medium. The advantages of steam were its availability, the low cost of water and minimal environmental effect. Heat transfer by steam uses latent heat. The saturation pressure dictates the temperature at which the heat transfer takes place. For high temperatures, the steam pressure has to be high. For example, to achieve a temperature of 350°C (662°F), a minimum pressure of 180 bars is required. This will need thicker and heavier steel for the heat exchanger tubes. This increases both weight and thermal stresses and will require special manufacturing techniques. All this leads to higher costs.

On this point, thermal oil scores the higher mark. Even at temperatures of 350°C (662°F), the pressure requirements are just sufficient to overcome the system pressure drops. This also decreases the pumping cost. The system is simple in that it requires only a pump, an expansion and storage tank and the heat exchangers. A steam system on the other hand requires demineralised makeup water supplies, drains, traps, safety valves, chemical additions and blowdowns. Using thermal oils eliminates all these complications together with issues of corrosion, scaling, fouling and the buildup of deposits in the heat transfer area. It is for this reason that thermal oil has become the preferred medium for heating. As an additional benefit, unlike steam, thermal oils can also be used in applications where temperatures are below 0°C (32°F).

Thermal oil heaters are functionally similar to steam boilers. The combustion chambers that burn fuel oil, biofuel, coal or any other fuel with all necessary safety devices are the same. As there is no evaporation involved, the oil passes through simple heater coils which are placed in the radiant or convection zones of the tank. The oil which is pumped through these coils heats up and flows to the process heat exchanger. The cooler oil returns to a tank and then back to the pump. An expansion tank takes care of the thermal expansion. The tanks have provisions to prevent the oxidation and vaporisation of the oil. Marine applications use very compact heaters with helical coils. Heaters downstream of gas turbines or diesel engines can additionally function as heat recovery systems for use in downstream processes adding to their cost effectiveness and efficiency.

Thermal oil is an interesting product which has a wide range of industrial applications. It has been widely employed by the oil, gas and chemical industry for the past 30 years and is an integral part of many chemical reactions. Storage and transportation of items such as asphalt, which have low solidification temperatures, use thermal oil heaters. This is especially suitable for when these products have to be transported by sea. Solar thermal systems or concentrating solar power (CSP) systems use thermal oils as a heat storage medium. This allows the power plants to produce power even when there is no sunshine. Heating biodiesel for transesterification, paper and board manufacturing and noble metal extraction also make use of thermal oils.

As may be evident, thermal oils are available in a wide range of specifications depending on their intended application and use. Different manufacturers produce different oils under different brand names. It is important to recognise how different specifications suit various needs of the process and applications. Some of the most commonly used brands of thermal oil are:

- Therminol from Solutia Inc.,
- Dowtherm from the Dow Chemical Company,
- Exceltherm from Radco Industries Inc., and
- Paratherm from Paratherm Corporation.

The most important characteristic is the maximum temperature of service. The oil is thermally stable up to this temperature. It should be borne in mind that the cost increases exponentially with each increase in the maximum temperature limit. Currently, thermal oils are available up to 400°C (752°F). Some oils are also available for cryogenic applications up to −100°C (−148°F). Apart from the temperature, other characteristics that determine which type or brand of thermal oil should be used include the heat transfer coefficient, pumpability and serviceability, environmental issues including toxicity, shipping restrictions and disposal methods, rate of oxidation and degradation potential. As with any system, thermal oil heaters have their foibles which the ship's engineers must beware of, including sudden trips or unplanned shutdowns of the system may cause the thermal oil to overheat; overheating causes degradation of the oil, resulting in the formation of sludge. This will necessitate the cleaning of the tank, sump and pipework/lines and the replenishing of the oil. Last of all, leaks, especially in the combustion area, can cause significant fire hazards. To avoid maloperation, it is essential that the valves, gaskets and packings are suitable and rated accordingly for thermal oil use. Piping and support design should account for thermal expansion and thermal fatigue. This issue is sometimes overlooked due to the low-pressure nature of the application. Despite these problems, manufacturers are constantly developing and producing thermal oils with higher temperature limits and specifications. Moreover, the industry is always finding new and novel applications for thermal oils, which in turn helps to improve energy efficiency and drive cost efficiencies in the maritime industry.

Chapter 9

Tank cleaning and gas-freeing operations

GENERAL

It is understood that tank cleaning is an essential task for any tanker. Even so, it must be recognised that it is a highly hazardous operation, and rigorous precautions must be observed throughout the process. Together with gas-freeing, tank cleaning is without doubt the most hazardous operation routinely undertaken onboard. The additional risk created by cargo gases expelled from the tanks cannot be overemphasised nor should they be underestimated. Depending on the most recent cargo carried in the tanks that are to be cleaned, vapours that are toxic, flammable and corrosive should be expected to be released into and around the cargo deck area. It is therefore of utmost importance that every precaution is exercised during all operations connected with tank cleaning and gas-freeing, and that the operations are carried out using only approved procedures and arrangements for the ship. Personnel involved in tank cleaning and gas-freeing should be fully aware of the dangers and take necessary precautions, as the consequences of an inadvertent error can be profoundly serious and far-reaching.

REQUIREMENTS FOR CARRYING OUT TANK CLEANING OPERATIONS

Tank cleaning should be carried out in accordance with the instructions provided by the vessel charterer or the Company and always under the supervision of the Master. Whenever a grade of cargo different from the previous one is to be loaded, and the cargo may be seriously affected if contaminated with the residue of the previous cargo, or when edible oils are to be loaded having carried toxic and hazardous cargoes, thorough tank cleaning should be carried out. When the tanks are to be gas-freed to check for the condition of the tank, or to conduct inspections or effect repairs on the internal structure of the tank at sea or in drydock, again, the tanks must

 DOI: 10.1201/9781003505044-9

be thoroughly cleaned. Tanks must also be cleaned thoroughly when the vessel is expected to take on clean ballast.

Cessation of tank cleaning operations

In certain circumstances it may not be advisable, or permissible, to clean the ship's tanks with seawater. This is usually the case during cargo handling operations, and/or when work is being done inside a tank, the cleaning of tanks adjoining that tank should, in principle, be prohibited. However, the mucking operation (removal of sludge) before drydocking and emergency cases in which the Master considers tank cleaning inevitable, may be excluded provided that appropriate safety measures are taken. The cleaning of tanks adjoining a tank filled with cargo liquid should, in principle, be prohibited. This does not apply to cases when the Master considers tank cleaning necessary. When a slop disposal ship or a barge is alongside the vessel for the purpose of receiving sludge, tank cleaning operations must cease. In any other situations which the Master considers dangerous, all tank cleaning operations must stop.

Suspension of tank cleaning operations

In the following situations, tank cleaning operations should be suspended on the Master's command: whenever weather and/or sea conditions will render the tank cleaning operation too hazardous to continue; when fire has broken out on or near the vessel; when another vessel is approaching and there is a risk of collision with that vessel; when flammable or toxic vapours have accumulated, unusually on deck; when there is a likelihood of thunder in the vicinity of the vessel; when it is difficult to drain tank cleaning water because of the ship's motions; when portable machines may come in contact with structures in the tank during tank cleaning operation due to the ship's rolling; when the oxygen content in the tank containing flammable vapours exceeds 8% during tank cleaning operations or whenever there is an inter gas plant failure; and ultimately in any other situation or circumstance when the Master considers it too dangerous. In cases where the vessel is not fitted with an inert gas system, it is recommended that tank cleaning should be carried out in a controlled atmosphere. The atmosphere of the tank should be frequently checked to ensure the tank is kept within the controlled atmosphere parameters.

All prewashing requirements for the crude and chemical cargoes must be complied with; it is strongly recommended that the ship's officers consult the International Bulk Chemical Code (IBC Code) or International Code for the Construction and Equipment of Ships carrying Dangerous Chemicals in Bulk (BCH Code) as and when such cargoes are carried.

RESPONSIBILITY AND SUPERVISION

Responsibility of the Master

The Master has the entire responsibility for the safe management and operation of the vessel under their command. That said, the actual operation of tank cleaning is usually carried out under the full charge of the Chief Officer, having sought approval from the Master. While the Chief Officer must maintain an up-to-date plan for managing cargoes taking account of all safety aspects of the last cargo carried in each tank and the requirements for the next cargo to be loaded, this is done under the authority and command of the Master. Tank cleaning is a critical activity and must be carried out with absolute regard for safety.

Responsibility of the Chief Officer

The Chief Officer must make a tank cleaning and gas-freeing plan in accordance with the P&A Manual as prepared by the shipyard. This manual will be approved by the Class and provided onboard for the approval of the Master. When making the plan, the Chief Officer should consider safety, efficiency and adjustment of the ship's condition in relation to retained ballast and slops in order to keep hull stresses within permissible limits and to ensure proper draft and trim throughout all stages of the operation. The P&A Manual should be consulted, as necessary. Before the commencement of operations, the Chief Officer must make known, in writing or verbally, the objectives and content of the operation, the procedures to be followed during each operation and the precautions to be taken. Furthermore, the Chief Officer must check and ensure the proper functioning of detectors, protective outfits, safety equipment and all other tools necessary for the safe performance of the task.

Responsibility of the duty officer

The duty officer should be completely familiar with the ship's various equipment and pipeline systems that are related to tank cleaning and gas-freeing operations. The duty officer must supervise the operation always in accordance with the instructions of the Chief Officer. The duty officer should make efforts towards safe and efficient tank cleaning operations, by maintaining close contact with the duty engineer in the engine department, to ensure all equipment and ship functions are operating accordingly.

PROCEDURES FOR TANK CLEANING OPERATION

The tank cleaning of crude and chemical tankers must be carried out in accordance with the procedures set out in the P&A Manual. For guidance, unless stipulated otherwise, this usually includes all or some of the following steps.

Tank cleaning methods

There are two primary methods for cleaning tanks. The first method involves washing by machines. In this context, tanks are washed using Butterworth machines. The second method involves washing by use of tank cleaning chemicals which are mixed with either warm or cold freshwater. Where tank cleaning chemicals are used, it is important to factor in any limitations that may preclude the use of specific cleaning agents. The main point of concern is how the cleaning agent may react to the presence of cargo residues inside the tank (refer to Figure 9.1).

Cleaning chemicals to be used in tank cleaning

Depending on the cargo carried, there are various types of cleaning agents and chemicals which can be used. As stated above, this will depend on the reactive nature of the chemical in relation to the cargo residues within the tank. The first type of cleaning agent is a solvent or emulsifier. These are solvents of aromatic hydrocarbons or aliphatic hydrocarbons. Generally speaking, aromatised hydrocarbons have strong cleaning powers but also have a strong smell, therefore adequate ventilation of tanks is necessary after cleaning to ensure future cargoes are not affected. Aliphatic hydrocarbons are effective against low-viscous cargoes as they have no smell but do physically mix with some cargoes as an emulsifier. Concentrations of solvents are usually mixed at a ratio of 2%–6%. The second type of cleaning agents are pure liquid emulsifiers. These are diluted with water into concentrations of 0.1%–0.05%. The concentration is then heated to between 60°C and 80°C (140°F–176°F) and applied with a pressure of between 7 kilograms and 10 kilograms per square metre. The third cleaning agent consists of alkaline detergents (light duty). These are stored onboard as either a powder or a liquid and are diluted with water when it is necessary to scrub down light residues or deposits that remain in the tanks. Alternatively, alkaline detergents (heavy duty) may be used where light-duty alkaline detergents will be ineffective. This usually applies when it is necessary to turn non-drying fatty acids water-soluble. The difference between light-duty and heavy-duty alkaline detergents is the presence of caustic soda and caustic kalium in heavy-duty detergents. Where it is inappropriate or potentially damaging to use chemical cleaners, neutral detergents may be employed instead. These consist of a combination of non-ion/anionic surface-active agents and glycol ether. Zinc-painted tanks must normally be cleaned using neutral detergents to protect the integrity of the zinc coating.

Figure 9.1 Tank cleaning equipment. (a) COW hose head and (b) COW appliances.

Immediately after the cleaning operation is concluded, the tank must be rinsed in order to wash away any remaining cleaning chemicals on the tank walls. Specially designed tank cleaning machines and brushes may then be used to remove any remaining content on the tank floor using warm/cold freshwater for about 2 hours. The final stage of the cleaning operation involves removing any remaining residues from the internal structures of the tank with freshwater. It is important to ensure all residues on the bottom of the tank and in the pipelines are completely flushed.

Steaming

Steaming involves the introduction of saturated steam into the tank, which condenses on the tank surface. The purpose of introducing steam to the tank is to (1) evaporate the remaining volatile residues (i.e., to remove residual smells) and (2) to reduce the presence of chloride. If it is intended to evaporate residues, it is mostly desired to raise the temperature as high as possible during steaming. This can be enhanced if the adjacent tanks (including ballast tanks) are empty. If the freshwater has elevated chloride levels, the use of steam for the removal of chloride is often the only feasible option albeit the steam quality will depend on the construction of the boiler. If the steam has low chloride levels, steaming is an effective method for reducing chloride levels. If the purpose of steaming is to remove chlorides, the tank wall temperatures should be cool in contrary to the evaporation method described above. This results in increased condensation and water film running down the tank walls which washes the chloride off the tank surface. Steaming may be carried out, provided there are no flammable vapours present in the tank and all associated pipeline systems, after completion of cleaning and gas-freeing of tanks and pipelines.

> Steam **MUST NEVER** be injected into a tank adjacent to any tank containing heat-sensitive cargoes.
>
> The steaming of tanks with toluene or methanol is prohibited.

Draining and drying

Tanks must be thoroughly ventilated and dried using fixed or portable blowing machines. During this operation, attention should be paid to preventing rainwater or seawater from entering the tanks.

PREPARATIONS FOR TANK CLEANING

Precleaning conference

A precleaning conference under the leadership of the responsible officer should be held prior to any tank cleaning or gas-freeing operation. Other crew members involved should be identified by the responsible officer and their roles explained. The conference should confirm:

- The tanks to be cleaned and the cleaning sequence,
- The type of cargo to be cleaned from each tank and its characteristics. Cargo information sheets should be available so that personnel involved are familiar with the hazards,
- The major risks during cleaning such as toxicity, flammability, corrosiveness and reactivity,
- The safety equipment and personal protective equipment to be available and ready for use throughout the operation and during connecting and disconnecting of hoses at the cargo manifold,
- The cleaning instructions to be followed in each case,
- The means of disposal of any cargo residues and the contaminated cleaning water. The relevant slop tank must be specified in each case,
- The precautions necessary to confirm that the cargo deck area is free from cargo vapours during tank washing and gas-freeing operations, and
- That at regular intervals throughout the operation, checks will be made to ensure that tank washings containing cargo are not inadvertently being discharged into the sea.

Preparations

A written tank cleaning schedule should be drawn up and made available for reference by all personnel participating in the tank cleaning and follow-up operation. Before any tank cleaning or gas-freeing operations begin, the responsible officer should confirm that all necessary equipment is available and that adequate checks are made to establish that all equipment to be used is in good working condition. Both before and during tank cleaning and gas-freeing operation, the responsible officer should be notified that the appropriate precautions set out in this chapter are being observed. All personnel on board should be notified that tank cleaning or gas-freeing is about to begin, and only the personnel involved in the operations should be allowed into the cargo tank area. If other craft are alongside the tanker, their personnel should be notified that tank cleaning operations are about to commence, and their compliance with all appropriate safety measures should be confirmed. When gas-freeing or tank cleaning while alongside at a terminal, the precautions for cargo handling be observed where appropriate. Before starting, the permission of the port authority and terminal operator should be obtained and the appropriate personnel ashore should be consulted to confirm that conditions on the jetty do not present a hazard and to obtain agreement that operations can start. The following checks should be made before operations commence:

- That essential protective clothing and respiratory protection equipment are being worn if so required,
- That freshwater shower and eyewash arrangements are ready for immediate use in the event of contamination of personnel,
- That work not related to cargo operations and not otherwise essential, is avoided in the cargo area during tank cleaning operations,
- That cargo pipelines serving a set of cargo tanks are isolated from the tanks to be cleaned or gas-freed unless all tanks in that set are to be cleaned,
- That tanks served by a common vent system are properly isolated,
- Those cargo tank lids, tank washing openings, ullage openings and sighting ports in uncleaned tanks are kept closed until they are to be cleaned,
- That all sea and overboard discharge valves connected to the cargo and ballast systems are shut and secured when not in use,
- That pump room precautions are being observed and will continue to be observed throughout tank cleaning and gas-freeing operations, and
- That firefighting equipment is ready for immediate use.

Use of water as a cleaning agent

Water is the most common agent used for washing and flushing cargo tanks. This is because water is readily available in large quantities and is an efficient cleanser. It has the added benefit of being easy to heat when necessary. Nevertheless, it is sometimes necessary to use small quantities of chemical additives or detergents as a cleaning agent in order to improve the cleaning effect. In some situations, water cannot be used lest it reacts violently with the cargo residues left in the tank. Where water cannot be used as a cleaning agent, it may be possible to use ventilation instead to remove cargo residues and gas-free the cargo tank after a highly volatile cargo has been carried. In every case, the full safety aspects of the operation should be considered. When tank cleaning in port, the regulations and limitations set by the port authorities must be complied with. After carrying a low flash point cargo, a flammable vapour mixture should always be suspected until tests have established that the atmosphere is non-flammable. Equally, care is necessary after carrying non-volatile flammable cargo at temperatures above its flash point or after discharging any cargo or ballast that had been loaded into a tank that was not free of flammable vapour. Toxic vapour in harmful concentrations should also be assumed after unloading cargoes which have a vapour inhalation hazard. The presence of cargo vapour – whether toxic or flammable – should be suspected in confined spaces such as cofferdams whenever there is a risk that cargo may have leaked into that space.

For tank cleaning operations, tank cleaning machines are usually used although it may be necessary to complete the tank cleaning with a manual handwash. For the effective cleaning of the tank bottom, the tank should be drained thoroughly. During tank cleaning operations, all precautions should be taken.

Hazards when using recirculated wash water and/or detergents

Recirculated wash water should not be used when the tank is being washed in an undefined atmosphere as it may increase the generation of static electricity. Cleaning by recirculation may be dangerous if the dirty oil/water mixture contains sufficient cargo of such a nature, that it will generate an electrostatic charge. Recirculation should not be practised unless the tank is free of flammable vapour or is inert. It may also pose a hazard when slops from cargoes which react dangerously together are discharged into the same recirculation tank. Both International Safety Guide for Oil Tankers and Terminals (ISGOTT) and United States Coast Guard (USCG) provide guidance on reactivity; however, care must be taken to avoid the introduction of unknown mixtures into the slop tank. If possible, commingling of slops should be avoided. When using a coated tank as a recirculation tank, the effect of the slops upon the coatings must be accounted for. Before commencing the recirculation, the integrity of all tank cleaning hoses and fittings must be checked under pressure. This should be done with water only. After the system has been confirmed tight, cleaning detergents may be added. The Butterworth lid should be covered with canvas to prevent cleaning water with detergent spraying on the deck. If tank cleaning chemicals are to be used it is important to recognise that certain products may introduce a toxicity or flammability hazard. Personnel should therefore be made aware of the threshold limit value (TLV) of each specific product. Detector tubes are particularly useful for identifying the presence of specific gases and vapours in tanks. Tank cleaning chemicals capable of producing a flammable atmosphere should only be used when the tank is certified clean and inerted.

The selection of detergents should be considered carefully, considering any toxic vapours that may be released through their introduction to the tank. Some products may be used for the local cleaning of tank bulkheads and blind spots by hand wiping, provided the amount of tank cleaning chemical is small and the personnel entering the tank observe enclosed space entry requirements. A thorough risk assessment should be carried out before any local cleaning is authorised. In addition to the above, the manufacturer's instructions and/or recommendations for the use of cleaning products must be observed. Where these operations take place in port, the Post State authorities may impose additional requirements. Material Safety Data Sheet (MSDS) for tank cleaning agents should be kept onboard for reference.

Tank washing atmospheres

Tank washing should only be carried out in an inert atmosphere. This means flammable vapours are rendered unhostile by the introduction of inert gas. The inert gas reduces the overall oxygen content within the tank meaning any vapours cannot burn. The oxygen content of an inert tank atmosphere should not exceed 8% by volume. Although the atmosphere in a properly inerted tank is incapable of burning, it is important that the following precautions are observed:

- When portable washing machines are used, all hose connections should be made up before the washing machine is introduced into the tank. Connections should not be broken until after the machine has been removed from the tank. However, to allow for the draining of the hose, a coupling may be partially opened and then retightened before the machine is removed,
- The tank should be kept drained during washing. Washing should be stopped to clear any build-up of wash water.

When there is a need to maintain an inert atmosphere during tank washing, the following points should be observed:

- The purity and pressure of the inert gas being delivered during the washing process should be proactively monitored,
- Before each tank is washed, the oxygen level in the tank should be determined at a point of 1 metre (3.2 feet) below the deck and at the middle level of the ullage space. At neither location should the oxygen level exceed 8% by volume, and
- If during the washing the oxygen level in the inert gas supply exceeds 8% by volume, or the pressure of the atmosphere in the tank is no longer positive, washing should cease until satisfactory conditions are restored.

It should be noted that some cargoes are carried under an inert blanket. Even though these cargoes pose no flammable hazard, the atmosphere is inerted simply to maintain the quality of the cargo. In these situations, it is important to treat the cargo space in the same manner as if the cargo does pose a flammable hazard.

Precautions when tank washing in an undefined atmosphere

Most tank cleaning on chemical tankers is conducted in an undefined atmosphere. In all cases after carrying flammable cargo, the atmosphere in an empty tank should be treated as flammable. The only way to guarantee that an explosion cannot occur during washing in an undefined atmosphere is to make certain that there can be no source of ignition either inside the tank or within proximity to the tank. Good tanker practice will avoid all normal sources of potential ignition but, even so, certain precautions should be taken, especially when there is a heightened risk from static electricity. Accordingly, before washing, the tank bottom should be flushed with water and stripped. The piping system, including cargo pumps, crossovers and discharge lines, should also be flushed with water. The flushing water should be drained to the tank designated to receive slops. This operation may not be necessary if the ship is fitted with an efficient stripping system, and the cargo tank and pipelines have been stripped as detailed in the ship's P&A Manual.

Note: if cargoes are highly water reactive, this operation must not be carried out.

When portable washing machines are used, all hose connections should be made up before the washing machine is introduced into the tank. Connections should not be broken until after the machine has been removed from the tank. However, to allow draining of a hose, a coupling may be partially opened and then retightened before the machine is removed. Ropes made of synthetic fibres should not be used to support tank cleaning machines as these attract static electricity. No machine should have a throughput greater than 60 m^3 per hour and no nozzle may have a throughput greater than 17.5 m^3 per hour, and the total water throughput per cargo tank should be kept as low as practicable and must not exceed 110 m^3 per hour. The tank should be kept drained during washing; washing should be stopped to clear any build-up of water inside the tank. Recirculated wash water should not be used as it may increase the generation of static electricity. Sounding rods and other equipment must be introduced through a sounding pipe reaching close to the bottom of the tank and earthed to it. If a sounding pipe is not used, then additional precautions should be followed. No material that has the potential to spark or cause static electricity should be lowered into the tank. Last of all, under no circumstances should steam be injected into the tank. Because of the hazard from static electricity, steam should not be introduced into cargo tanks where there is a risk of the presence of a flammable atmosphere. It should be borne in mind that a non-flammable atmosphere cannot be guaranteed in all cases where steaming might be considered.

Precautions for sounding tanks when not using a sounding pipe

If a sounding pipe is not used, it is essential that any metallic components of the sounding rod or other equipment are bonded and securely earthed until fully removed from the tank. This precaution should be observed during washing and for a minimum of 5 hours afterwards unless the tank is continuously and mechanically ventilated after washing. In the case of the latter, the delay period can be reduced to no less than 1 hour. During the delay period, an interface detector of metallic construction may be used if earthed to the ship by means of a clamp or bolted metallic lug. A metal rod may be used on the end of a metal tape which is earthed to the ship. Metal-sounding rods suspended on natural fibre rope should not be used even if the end at deck level is fastened to the ship. This is because the rope cannot be completely relied upon to act as an earthing path. Alternatively, equipment made entirely of non-metallic materials is considered safe to use, for example, a wooden-sounding rod or float suspended on a rope would not require earthing.

Under no circumstances should ropes made of synthetic polymers or metallic chains be used for lowering equipment into cargo tanks.

Cleaning of cofferdams or double-bottom tanks

If it is necessary to clean cofferdams or double-bottom tanks into which cargo liquids or vapour could have leaked, the same precautions should be observed as when cleaning cargo tanks.

Freefall of wash water in slop tanks

It is essential to avoid the free fall of slops or tank washing water in a receiving slop tank unless the tank is inert. Washing water or slops should be transferred to the receiving tank through the cargo system. If a different arrangement is necessary, then to avoid splashing the receiving tank should be filled to a depth of at least 1 metre (3.2 feet), or sufficient to ensure the discharge inlet is well below the surface of water.

Special cleaning methods

Water washing may be inadequate or inappropriate following the carriage of certain products as the tanks can only be adequately cleaned using special cleaning methods or cleaning agents. Where it is appropriate to use a special cleaning method, always follow the manufacturer's guidance. Where any special cleaning method is to be used in the port, local authorities may impose additional safety or environmental requirements. Some cargoes may react with certain cleaning agents and produce large amounts of toxic or flammable vapours or render equipment such as pumps inoperable. The choice of tank cleaning agent should be made only after consultation and in full knowledge of the cargo's characteristics. Furthermore, special methods involving cleaning chemicals and additives may create additional hazards for the crew. The ship's staff must therefore be made familiar with the hazards involved and the necessary precautions required. Personal protective equipment (PPE) is to be made available to all personnel involved in the cleaning operation. Cleaning agents may be added to wash water or used alone. Whichever method or agent is chosen, the cleaning procedures adopted should not entail the need for personnel to physically enter the tank. If, however, the only practical means of cleaning involves personnel entering the tank, then the precautions mentioned previously are to be strictly followed. No one is to be authorised or allowed to enter the tank until all checks have been satisfactorily made as per the Enclosed Space Entry Permit contained in the Ship's Safety Manual and permission has been obtained from the Master. Chemical absorption detectors may be used for detecting the presence of specified gases and vapours at specified TLV levels.

In exceptional circumstances, a requirement might arise for wiping down product residues from the tank walls using a chemical solvent in a localised area. In these cases, the amount used should be small, and the personnel involved should be made aware that its use may modify the atmosphere inside the tank. The introduction of the solvent into the tank might also generate additional risks such as toxicity or flammability. Such risks should be carefully evaluated before starting the operation, which should not be undertaken unless the personnel involved can be effectively protected from the risks identified. The MSDS for the chemical solvent(s) to be used must be available onboard and referred to for guidance relating to potential hazards and personnel protection measures. A detailed risk assessment is to be carried out and approved by the Master prior to carrying out the cleaning operation. In addition, the manufacturer's instructions or recommendations for the use of commercial products should be observed, and the resulting slops disposed of in accordance with the ship's P&A Manual (refer to Figure 9.2).

For guidance relating to the use of chemicals and additives, refer to the latest MEPC.2/Circ.19/Annex 10 (cargo tank cleaning additives evaluated in accordance with MEPC.1/CIRC.590, which are certified compliant with the requirements of regulation 13.5.2, Annex II, MARPOL 73/78).

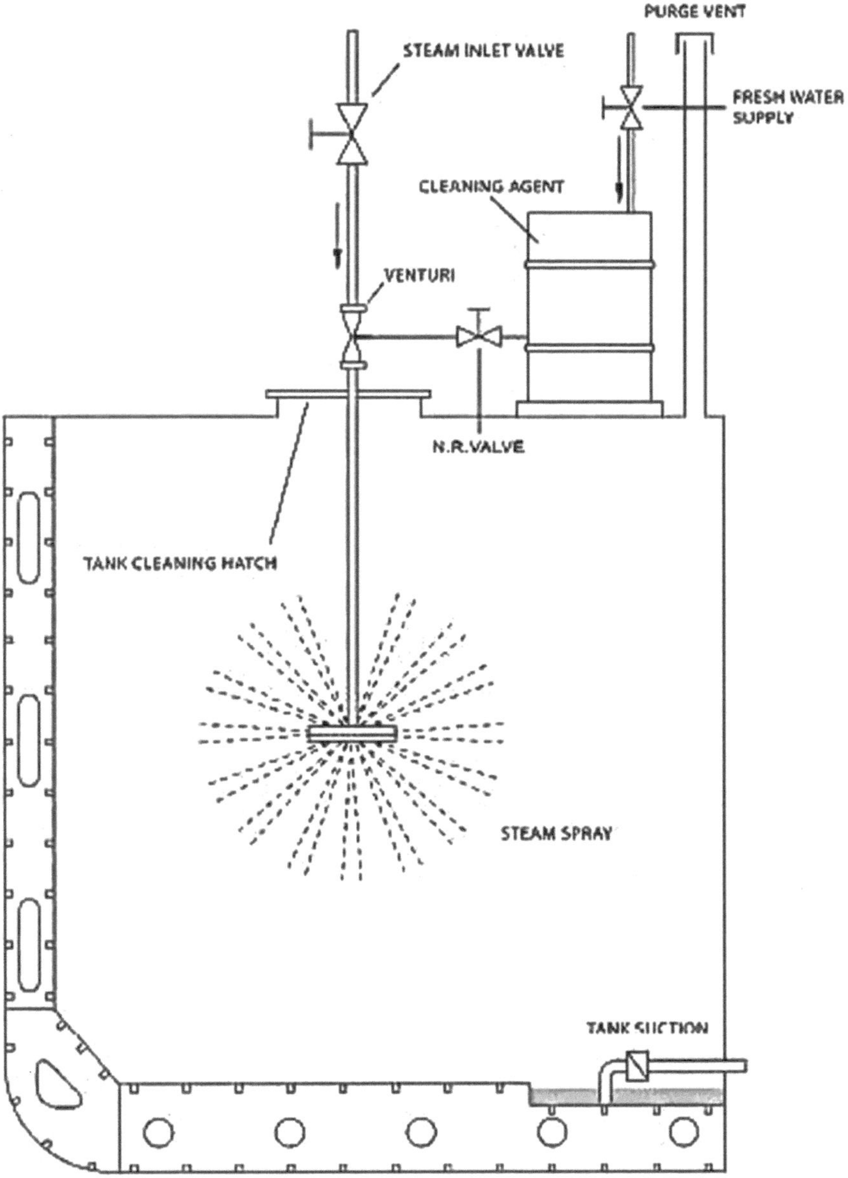

Figure 9.2 Modern chemical tanker tank cleaning process using steam spray.

Indicative cleaning methods for various types of cargoes

For instructions relating to cleaning methods for specific types of cargoes refer to the latest version of the Tanker Safety Guide (Chemicals), which is provided onboard for tank cleaning. In addition, the vessel must also demonstrate compliance with any specific tank cleaning requirements mandated by the shipper's terms and conditions. If there is any doubt regarding the use of cleaning procedures or the agents to be used, the vessel must liaise directly with the shipper and/or the cargo manufacturer. It is vital never to assume that a cleaning method or agent is acceptable based on prior use. The steaming of tanks with toluene or methanol is prohibited.

While the following points are provided for guidance, it is the responsibility of the vessel to confirm the specific measures to be taken in preparation for, during and after, tank cleaning.

1. Lubricating oils and additives. Firstly, tanks should be cleaned with water using the fixed tank cleaning machines. Then as required, cleaning by detergents and steaming can commence.
2. Middle and high-grade alcohols. As high-grade alcohols have a high solidifying point, therefore cleaning with warm water is usually necessary in winter months. After cleaning with chemical solvents or detergents and deodourising by steaming, tank cleaning with fixed cleaning machines can commence.
3. Acrylic ethers. Tanks after the discharge of cargo should be gas-freed first as these cargoes have a peculiar smell. The tanks can then be washed with cold water since these cargoes polymerise when subject to light and/or heat. The tanks can then be washed with fixed cleaning machines after cleaning with detergents.
4. Chlorides. Chlorides are a special kind of solvent as they are extremely volatile but are insoluble in water. Where a tank had previously contained chlorides, the tank must be washed own thoroughly to ensure there are no remaining residues. Both organic and inorganic chlorides are likely to cause damage to the plant and ship's structure if left unattended. To prevent this, the tanks must be gas-freed after the discharge of cargo and washed with fixed cleaning machines. If necessary, additional cleaning using detergents and steaming may be necessary to remove stubborn residues.
5. Resin. Following discharge, the tank must be gas-freed as resin cargoes are volatile and toxic. The tanks should then be washed with fixed cleaning machines and steamed.

6. Phenols. Tanks containing phenols such as cresols, etc., should be adequately washed with water and detergents/solvents and then steamed. The tanks should then be washed with fixed cleaning machines. Nonylphenol is extremely viscous and not soluble even in hot water, therefore specialised detergents and/or solvents will be required.
7. Animal, vegetable and fish oils and fats.
   a. Drying oils (linseed oil, soya bean oil, sunflower oil etc.) Immediately after the discharge of these cargoes, the tanks should be washed with warm water (~30°C–35°C) (~86°F–95°F) in the following manner:
      i. On the upper and middle parts of the tanks: 30 minutes,
      ii. On the lower and bottom parts of the tanks: >1 hour.

      The tanks must then be inspected. If any residues are found, these must be removed manually using light-duty alkaline detergents and portable washing machines. The tank should then be rinsed with seawater. If it is not possible to clean down the tanks immediately after discharge, heavy-duty alkaline agents must be sprayed into the tanks and left sealed for a minimum of 24 hours. Under no circumstances should anyone enter the tank before at least 24 hours have passed.
   b. Semi-drying oils (corn/maize oil, cottonseed oil, fish oil). Cleaning should be carried out in accordance with the method for drying oils; however, it is preferable to use softer detergents than those used for drying oils.
   c. Non-drying oils (coconut oil, peanut oil, cod liver oil, rapeseed oil, palm oil, olive oil, whale oil etc.). Immediately after discharge, the tanks must be washed down with hot water (minimum temperature: 80°C (176°F) at a rated pressure of 10–12 kg/cm^2, in accordance with the following:
      i. On the upper and middle parts of the tank: >1 hour,
      ii. On the middle and bottom parts of the tank: >1.5 hours.

      Thereafter, cleaning with detergents or solvents may be done. Always check the P&A Manual for the acceptable temperature ranges with respect to tank coatings and valve seat rings; the maximum operating temperature that the vessel is designed for must never be exceeded.
8. Molasses. Immediately after the discharge of cargo, the tanks should be washed with hot water (minimum temperature: 80°C (176°F) at a rated pressure of 10–12 kg/cm^2, in accordance with the following:
   i. On the upper and middle parts of the tank: ~1 hour,
   ii. On the lower and bottom parts of the tank: ~2 hours.

   An alternative method is to wash down the tanks with cold water on the upper, middle and lower parts for 30 minutes, then again

Figure 9.3 Tank cleaning heater.

wash with a warm 3% solution of heavy-duty alkaline detergents on the upper, middle and lower parts of the tank for 30 minutes. An internal inspection of the tanks should be carried out after each washing. If residues are still found inside the tanks, then these spots should be sprayed again with 20% solution of heavy-duty alkaline detergents or non-diluted light-duty alkaline detergents using a portable cleaning machine. The tank should be left for about 1 hour. After 1 hour has passed, the tanks can be rinsed with seawater and then freshwater. Always check the P&A Manual for the acceptable temperature ranges with respect to tank coatings and valve seat rings; the maximum designed operating temperature that the vessel is designed for must never be exceeded.

9. Monomers. As monomers are liable to polymerise under the influence of heat and/or light, tanks must be gas-freed first before being washed down with cold water and then cleaned with detergents and steaming. Once the tank has been rinsed, tank cleaning with fixed cleaning machines can commence (refer to Figure 9.3).

Gas-freeing operation

Execution of gas-freeing operation

Gas-freeing onboard tankers is required for entry into the cargo tanks, for hot works, and/or when commencing washing to clean the ballast tanks. Gas-freeing is one of the most hazardous operations routinely undertaken onboard tankers. Notwithstanding the procedure itself, additional risks are created by the cargo gases being expelled from the tanks. These may be toxic, flammable and corrosive. It is therefore extremely important that absolute care is exercised during gas-freeing operations. The consequences of an inadvertent error can be serious and have far-reaching consequences for both personnel and the environment. A space is considered "gas-free" when the concentration of flammable gases in its atmosphere is less than 0% lower explosive limit (LEL); the concentration of toxic gases (including inert gas components) is less than the TLV and the oxygen concentration is not less than 20.8%–21%. Given that hazards lurk at various stages in the gas-freeing process, the following recommendations should be followed. Furthermore, the IBC Code also contains expert advice regarding cargo tank gas-freeing. Primarily, it is essential to know what type of vapours can be expected and whether they are flammable and/or toxic and/or corrosive. Thereafter:

1. Venting of toxic and flammable gas during gas-freeing should be through the vessel's approved gas-freeing outlets, and therefore, the exit velocity should be sufficient to carry the vapours clear of the deck. No escape of cargo vapours should occur at deck level before the concentration within the tank has fallen below 30% lower flammable limit (LFL) and the relevant TLV. Final clearance of the vapour mixture may continue at tank deck level through other larger deck openings,
2. If portable ventilation equipment is to be used to blow air into a tank, tank openings should be kept closed until work on that tank is about to commence,
3. Where cargo tanks are gas-freed by means of permanently installed fans, air is introduced into the cargo tank through the cargo lines. The entire line system should be thoroughly drained before venting to avoid any obstruction of the airflow or tendency for water or cargo residues to be blown into a cargo tank. Valves on the systems, other than those required for ventilation, should be closed and secured. The fans should normally be blanked or disconnected from the cargo tank system when not in use,
4. Fixed gas-freeing equipment should not be used for gas-freeing of a tank while simultaneously being used to ventilate another tank in which washing is in progress, regardless of the capacity of the equipment,
5. Portable fans should only be used if they are water-driven or hydraulically or pneumatically driven. Their construction materials should be such that no hazard of incendiary sparking arises if, for any reason, the impeller touches the inside of the casing. The manufacturer's recommendations for maintenance should be followed. Guards should be in place to prevent accidental contact with fan blades,

6. Portable fans, where used, should be placed in such positions and the ventilation openings so arranged that all parts of the tank being ventilated are effectively and equally gas-freed. Fans should be located as remote as possible from the ventilation outlets,
7. Portable fans should be so connected to the deck that an effective electrical bond exists between the fan and the deck,
8. The wind direction may cause cargo vapours to pass near to air intakes for accommodation spaces or engine room ventilation and necessitate additional precautions. Central air conditioning or mechanical ventilation system intakes should be adjusted to prevent the entry of gas, if possible, by using recirculation of air within the spaces,
9. If at any time it is suspected that gas is being drawn into the accommodation block, the central air conditioning and any mechanical ventilating systems should be stopped, and the intakes covered or closed. It is unlikely that any ship now uses window-type air conditioning units which draw in air from outside the superstructure, but any which are still in use, or other plants which are not certified as safe for use in the presence of flammable gas, should be electrically disconnected and any external vents or intakes closed,
10. If the tanks are connected by a common venting system, each tank should be isolated to prevent the transfer of gas to or from other tanks,
11. When a tank appears to have been gas-freed and all mechanical ventilation has been stopped, a period of about ten minutes should elapse before taking final gas measurements. This allows (relatively) stable conditions to develop within the tank space. Tests should then be made at several levels and, where the tank is sub-divided by a wash bulkhead, in each compartment of the tank. In large compartments such tests should be made at widely separate positions. If satisfactory gas readings are not obtained, the tank should be checked for cargo residues and then ventilation resumed,
12. On completion of all gas-freeing and tank washing, the gas venting system should be carefully checked, with particular attention being paid to the efficient working of the P/V valves and any high-velocity vent valves. If the valves or vent risers are fitted with devices designed to prevent the passage of flame, these should also be checked and cleaned if found necessary. Gas vent risers and their drains should be checked to ensure that they are free of any blockage,
13. On completion of gas-freeing, attention should be given to all equipment that has been used and to enclosed or partially enclosed spaces that can retain or contain cargo residues or vapours, so that no unsuspected dangerous pockets can remain. Places where such cargo traces may exist include cargo lines, cargo valves, cargo pumps, stripping lines and valves, venting lines and P/V valves, vapour return lines, ullaging or sounding arrangements, heating coils, cargo

handling equipment storerooms, protective clothing storerooms and cargo sample storerooms,

a. The gas-freeing operation is carried out in the following cases:
   i. When a special type of cargo is to be loaded and its properties may be changed by the admixture of the previous cargo gas, instructions are given by the charterer on each occasion,
   ii. When persons will enter the tank, and
   iii. When drydocking.
b. Gas-freeing operation must not be carried out in the following cases:
   i. During cargo handling operations,
   ii. When another vessel is alongside the vessel. However, when a slop disposal ship or a barge is alongside the vessel for the purpose of receiving sludge before drydocking, these should be excluded,
   iii. When the local regulations or terminal regulations prohibit the operation, and
   iv. Any other situations which the Master considers dangerous.
c. The gas-freeing operation should be suspended in any of the following cases:
   i. When fire has broken out or on near the vessel,
   ii. When another vessel is approaching and there is a risk of collision with the vessel,

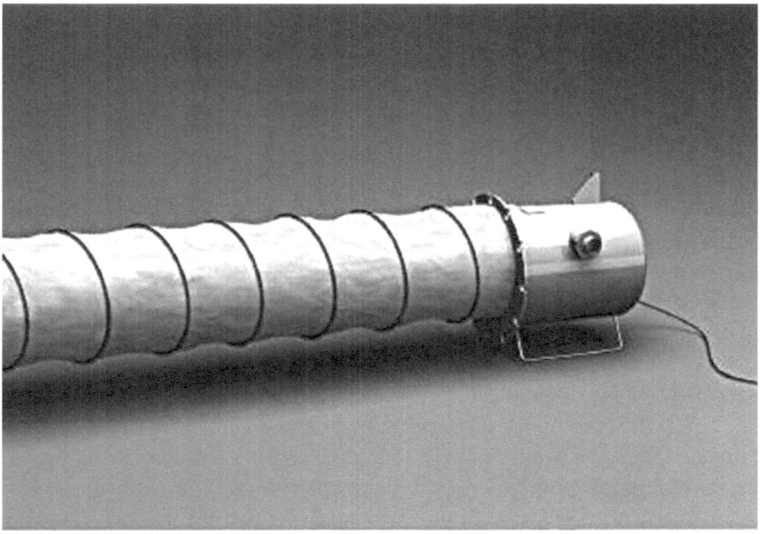

Figure 9.4 Chemical tanker tank ventilation hose.

iii. When there is thunder in the vicinity with the threat of lightning,
iv. When flammable/toxic gas has found its way into the living quarters and machinery spaces and it is impossible to prevent further such entry, and
v. Any other situations which the Master considers dangerous. (Refer to Figure 9.4.)

Procedures and precautions for gas-freeing operations

Always comply with the IBC/BCH Code and ICS Chemical Tanker Safety Guide requirement for gas-freeing crude and chemical cargoes (ref: IBC Code, ch. 8).

As mentioned previously, a space is considered as "gas-free" when the concentration of flammable gases in its atmosphere is 0% LEL, the concentration of toxic gases (including inert gas components) is less than the TLV and the oxygen concentration is not less than 20.8%–21%. The Chief Officer is responsible for supervising the gas-freeing operations. The gas-freeing programme and the progress of operation must always be available to all personnel involved. Protective clothing, resuscitation and firefighting equipment must be ready for immediate use. All doors, ports and windows are to be kept firmly secured. The inerted space can be purged with fresh air using the inert gas fans or water-driven portable fans. Careful readings of the tank atmospheres must be taken using the ship's portable gas detection equipment throughout the gas-freeing operation. A log of the readings must be maintained through the earlier stages of the operation. Hydrocarbon readings are to be taken with the Tankscope (i.e., hydrocarbons by volume) and during the completion of gas-freeing with the Explosimeter or Dräger Multigas Detector. Many vapours are heavier than air, and after they escape from the tank openings or vents, will tend to suspend around deck level. Where there is zero to light wind (i.e., below 5 knots) flammable or toxic mixtures may fail to disperse and instead lie about at some distance from where they arise. These gases can be carried through openings in the ship's superstructure to the galley, accommodation block, deck lockers etc. or else be drawn into machinery spaces. It should always be suspected, even after spaces have been cleaned and made gas-free, that some cargo liquid or vapour – or both – remains suspended or else may be released when pumps, cargo lines, valves, heating coils etc. are opened up. Appropriate precautions must be taken against such releases. Due to the risk of air pollution, gas-freeing operations must not be carried out within port limits without the express permission of the appropriate Port State and terminal authority.

When the vessel is not fitted with an inert gas system, the gas-freeing operation should be conducted such that the flammable vapour is discharged accordingly:

- Outlets at least 2 metres (6.5 feet) above the cargo tank deck level with a vertical efflux velocity of at least 30 metres/second maintained during the gas-freeing operation; or
- Outlets at least 2 metres (6.5 feet) above the cargo tank deck level with a vertical efflux velocity of at least 20 metres/second and which are protected by suitable devices to prevent the passage of flame, and
- The above outlets shall be located not less than 10 metres (32 feet), measured horizontally, from the nearest air intakes and openings to enclosed spaces containing a source of ignition and from deck machinery, which may include anchor windlass and chain locker openings, and equipment which may constitute an ignition hazard, and
- When the flammable vapour concentration at the outlet has been reduced to 25% of the LFL, gas-freeing may continue at the cargo tank deck level. Refer to SOLAS Chs. 2–5 and the IBC Code, ch. 8 for further details.

Gas-freeing precautions

1. Check what type of vapours are involved; they may be flammable, toxic, corrosive or a combination thereof,
2. Alert other non-essential crew, including the engine room, that gas-freeing is to take place and that non-essential personnel should stay clear of the deck areas,
3. Wind direction may cause vapours to enter accommodation spaces or the engine room, which may necessitate precautions (i.e., at sea the vessel may need to be turned into the wind),
4. Portable ventilation equipment must be checked and fully operational,
5. Personnel involved in gas-freeing must wear PPE as appropriate,
6. Vapour must only be released through openings as stipulated in SOLAS and the IBC/BCH Code,
7. After gas-freeing, no tank entry may be permitted until the Enclosed Space Entry Permit has been issued by the responsible officer and the tank has been "tagged" as safe to enter, and
8. All fixed and portable gas detection equipment must be in operation and suitably calibrated throughout the operations.

Some vessels are provided with a fixed gas-freeing system comprising a fan unit connected to a pipeline on deck. This pipeline might be the cargo lines, the vapour return line or the inert gas line. When gas-freeing operations are completed, the fan must be completely isolated from the pipeline either by a removable spool piece or a blanking arrangement. This is to ensure flammable or toxic vapours cannot reach the fan after the tank has been loaded. In these situations, the following guidelines should be adhered to:

- A responsible officer must supervise all gas-freeing operations,
- All personnel onboard should be notified that gas-freeing is about to begin,
- Strict "NO SMOKING" regulations should be implemented and enforced,
- Instruments to be used for gas measurement should be calibrated and tested in accordance with the manufacturer's instructions before starting operations,
- Sampling lines should, in all respects, be suitable for use with and impervious to the gases involved,
- Only portable fans or blowers which are approved for use in gas-hazardous areas are to be approved for use,
- Venting of toxic and flammable gas during gas-freeing should be through the vessel's approved gas-freeing outlets. No escape of cargo vapours should be allowed in the proximity of the vessel's accommodation spaces. The velocity of venting should be sufficient to carry the vapours clear of the vessel's deck,
- If gas-freeing is done through cargo pipelines, the entire line system should be thoroughly drained before venting. Gas vent risers drains should be cleared of water, rust and sediment and any steam smothering connections tested and proved satisfactory,
- When the gas level within the tank has fallen to below 25% of the LFL and below 50% of the TLV for known toxic gases, the other deck openings of the tank may be opened to complete ventilation,
- Intakes of central air conditioning or mechanical ventilating systems should be adjusted to prevent a vacuum from developing inside the vessel's accommodation. Window-type air conditioning units which are not certified as safe for use in the presence of flammable gas, or which draw in air from outside the ship's superstructure, must be electrically disconnected and any external vents and intakes closed. If at any time it is suspected that gas is being drawn into the accommodation, central air conditioning and mechanical ventilation systems should be stopped and the intakes covered or closed,

- Fixed gas-freeing equipment may be used to gas-free more than one tank simultaneously; however, gas-freeing with a fixed system must not be carried out while tank washing is in progress in any other tank,
- If several tanks are connected by a common venting system each tank should be isolated to prevent the transfer of gas to or from another tank, and
- If petroleum vapours persist on deck in high concentrations, gas-freeing should be stopped with immediate effect.

When a tank appears to have been completely gas-freed, a period of about 30 minutes should elapse before taking final gas measurements. This should allow for (relatively) stable conditions to develop within the tank space. The testing of the atmosphere should be carried out to determine its flammability, as a percentage of LFL, and its toxicity relative to the TLV. Tests should be made at several levels. The venting must be stopped while carrying out atmosphere checks inside the tanks.

Electrical storms

In the event electrical storms develop in the immediate vicinity of the ship, all cargo operations, gas-freeing and tank cleaning operations which involve flammable cargoes must be stopped.

Effect of wind

Most chemical vapours are heavier than air, therefore cargo vapours released during loading, gas-freeing or accidental spills, may concentrate around the lower areas on deck, especially in conditions where there is little to no wind. Strong winds may create low pressure on the lee side of deckhouses or other structures, thereby leading vapour to be carried in that direction. Personnel should be alert to either possibility. Vapour clouds may also develop over and above the vessel. If necessary, the vessel's course is to be altered to ensure a favourable wind direction while at sea; where this is not possible or safe, cargo/gas-freeing operations must cease until conditions are rendered safe.

Gas testing

After completing gas-freeing, gas detection in the tanks and on deck must be done in accordance with the vessel's standard operating procedure (SOP). The following considerations should also be considered.

Precautions for gas testing

For the measurement of the concentration of flammable/toxic gases and oxygen, should be paid attention to the following matters:

1. The measurement of vapours by gas testing devices should be carried out by the Chief Officer. In the event the Chief Officer cannot take measurements themself, this duty must be delegated to another deck officer, however, the Chief Officer should confirm the final measurements. The gas measurement results should be entered into the "Oxygen Content/Flammable Gas, etc. Measurement Record" and thereafter reported to the Master,
2. Gas testing should be conducted in such a manner as to take samples at the upper, middle and lower levels of the tank, and from as many points as possible. When the tank space is divided by a swash bulkhead, samples should be taken in each separate compartment,
3. In a large compartment, samples should be taken at various locations throughout the range of the compartment (as per [2] above),
4. The measuring device should stand at the position where crew members feel the wind from the port or starboard side to prevent inadvertent inhalation of toxic gas,
5. When work in a tank is intended, measurements should be taken locally on site, in addition to the measurements from on deck. When such local measurement is taken inside the tank, safety provisions should be ensured beforehand by measuring the atmosphere in the vicinity from the deck, by posting a lookout on deck and issuing/using a lifeline, as required, and
6. As the concentration of hydrocarbon gas may rise with the increase in air or seawater temperature, consideration should be given to the interval at which gas testing is conducted.

Handling gas testing devices

Gas testing devices should be inspected and maintained in accordance with the manufacturer's user manual. Before each use of the device, a zero-point adjustment should be made after flushing the detector and sampling the hose with clean air. The span adjustment should be made using a standard gas sample. The equipment must be calibrated at least every three months (or as directed by the ship's SMS, and prior to each entry into a compartment). User checks must ensure there are no leaks from the sampling hose or defective connections. Moreover, care should be taken to keep the tip of the sampling hose away from any water or oil that may be found at the bottom of the tanks or on other structures as this can impact the effectiveness of the sensors.

Working in enclosed spaces

While personnel are inside enclosed and/or confined spaces, it is critical that sufficient ventilation is provided, and the atmosphere is monitored at regular intervals. Frequent atmosphere tests, as per the enclosed space entry permit, must be carried out to ensure the atmosphere inside the space remains safe. Working in confined spaces requires a degree of good physical and mental health and is certainly not for the claustrophobic, those with heart or breathing problems or those with physical disabilities or limited mobility. The proper identification and elimination of potential hazards at the tank cleaning planning stage will certainly enhance the health, safety and environmental performance of the tank cleaning project. Therefore, a detailed risk assessment should be carried out in advance of such work, wherein all hazards associated with the job are identified and control measures implemented to mitigate/minimise risks to an acceptable level.

Hazards relating to tank cleaning are as myriad and varied as the designs and physical dimensions of the tanks themselves. Personnel entering tanks will inevitably become exposed to any number of physical hazards such as:

- Climatic conditions,
- Slippery surfaces, tripping and obstruction hazards,
- Hot cargo, hot surfaces and heating coils,
- Lack of oxygen/atmosphere toxicity,
- Hazards imposed due to the location and physical design of the tank,
- Inadequate lighting/illumination, and
- Hazards introduced by way of tank cleaning equipment and the method of cleaning.

Climatic conditions can create unsafe conditions for those engaged in the tank cleaning process. Excessive internal temperatures are particularly difficult to manage, and personnel must be allowed regular breaks for rest. Regular consumption of water is important for hydration; salt tablets may be consumed to replace salt loss through perspiration. Extreme cold and excessive noise can also be equally debilitating. Slippery surfaces, trip hazards and other obstructions proliferate throughout storage tanks; therefore, it is essential that personnel are provided training on how to move through tank structures and give insight into the internal tank layout prior to entry. Unfortunately, mobility is often hampered further due to poor lighting conditions, which are often difficult to improve due to the presence of flammable atmospheres and/or the physical dimensions of the tank. Falling objects, slippery surfaces, poor housekeeping, obstructions at head height in floating roof tanks, poorly maintained or inappropriate equipment etc. are other potential sources of danger which can be minimised by adequate risk assessment and training. When performing manual cleaning, many hazards may exist due to the location and design of the tank, and many more are introduced by way of the tank cleaning equipment and personnel themselves. Subsequently, adequate precautions must be taken prior to making entry into the enclosed space.

Wall wash procedure

This section describes the approved method for collecting and analysing wall wash samples to determine the presence of contaminants on the bulkheads. The procedure involves contacting a constant area of the bulkhead with a given amount of specification grade methanol, collecting the liquid and analysing it for the presence of chlorides, hydrocarbons, colour and particulate matter or whatever might be required by the Charterer.

Precautions

1. Safety Considerations – eye protection is required when collecting the samples to prevent the inadvertent contact of methanol with the eyes during the sample collection process. Gloves should be worn to prevent the absorption of methanol into the skin,
2. Disposable plastic gloves are also worn to prevent contamination of the samples during the collection process (a sufficient amount of chlorides can be absorbed from the skin to cause the sample to fail the chloride analysis),
3. Chlorides are abundant in the marine environment. All sampling equipment including bottles, funnels and other apparatus must be thoroughly rinsed with methanol (of less than 0.2 ppm chlorides content) and stored in plastic containers. Bottles are to be capped prior to sample collection,
4. Personnel collecting the samples must be certain that no perspiration or bare skin contacts the sample or sampling equipment while the wall washes are being collected,

As a minimum, four sites of approximately 1.2 square feet each must be chosen in each tank. If additional sites are chosen, 100 millimetres of methanol should be applied to each location and collected in a separate container. Any area that indicates the presence of crystalline deposits should definitely be tested. Separate tests of non-typical areas greater than 2 square feet (e.g., discoloured patches etc.) should be conducted. The sample collected should be labelled with a description of the non-typical area. These areas should be analysed separately.

Sample collection procedures

Choose a minimum of four surfaces to test:

1. Using the plastic wash bottle, squirt methanol on the test section at the highest practical point (normally between 1.5 metres and 2 metres (4.9 feet–6.5 feet) above the tank bottom in a stream of about 10 centimetres (0.32 inches) wide,
2. Allow the methanol to run down the wall by 15 centimetres (0.49 inches) and begin collecting it with the funnel, squirting additional liquid as necessary to rinse the flushing into the sample funnel,
3. Continue this process until an area of 10 × 120 centimetres (0.30 × 47 inches) has been rinsed with 100 millilitre of methanol,
4. After the washings from the four sites are collected, submit a portion of the sample for analysis of chlorides, colour, suspended matter and hydrocarbons, whatever is applicable. The accuracy of this test depends upon consistency, i.e.:
    a. A consistent number of sites are tested,
    b. A consistent area is tested at each site,
    c. A consistent amount of methanol is applied to each site, and
    d. A consistent amount of methanol is recovered from each site.

For the purpose of standardising methods and maintaining the desired consistency, the following may be followed:

1. Four wall wash sites.
    a. An area of 10 × 120 centimetres for each wall wash,
    b. 100 millimetres of methanol applied to each site, and
    c. 250 millimetres total minimum recovery of methanol (approximately 60% of each of the four 100 millimetres washings).
2. Equipment and reagents:
    a. Polyethylene washing bottles, 500 millimetres capacity,
    b. Bottles, glass with screw cap and polyethylene lined, of sufficient capacity to hold the washing samples,
    c. Plastic disposable gloves,
    d. Specification grade methanol (laboratory pure methanol) that has been tested to be less than 0.1-ppm chloride by ion chromatography (**note:** high-quality methanol is vital to the accuracy of this test), and
    e. Sample funnel, plastic or stainless steel with one flat side that can be held flush with the bulkheads.

Disposal of tank washings, slops and dirty ballast

During the normal operations of a tanker, the need to dispose of chemical residues, slops or water contaminated with cargo will arise during or immediately after each tank cleaning. Final disposal of slops or wash water should be in accordance with the ship's P&A Manual. When tank washing is carried out at sea, tank washings and slops may be retained onboard in a slop tank, discharged ashore or pumped into barges.

Mandatory prewash water

Mandatory prewash procedures should be conducted strictly in accordance with the ship's P&A Manual, and the resulting contaminated wash water should always be discharged to shore. The intention of MARPOL is that this should happen immediately following cargo discharge operations and in the same port. However, occasions do arise when adequate shore reception facilities for the washings are not provided, and the ship must retain the washings onboard until arrival at another port. MARPOL addresses this matter, and the P&A Manual will provide guidance on the correct procedures for a particular ship. During such a voyage, the slops and tank washings should be given the same safety and environmental care as the original cargo.

Dirty ballast

Dirty ballast, caused by ballasting into a cargo tank before the tank is cleaned, should be treated as slops and must be disposed of in accordance with MARPOL and the ship's P&A Manual.

Safety precautions during discharge of cargo slops into the sea

When discharge overboard is permitted, it should only be undertaken when the ship is at sea and normally be below the waterline through an underwater discharge outlet on the side of the ship. This outlet must be located as far away from essential water inlet valves as possible. In the interests of safety, this procedure should be adopted even when it is not a mandatory requirement. When any discharges are made above the waterline, care should be taken to avoid cargo vapour or liquid blowing back on board. If such a risk exists, discharge should be made below the waterline: if this is not possible, consideration should be given to altering the ship's course or speed to reduce the risk, and personnel on deck should wear appropriate protective clothing.

Management of slop tanks

Compatibility of various cargo and cleaning chemicals should be considered just as carefully when handling slops as when handling the cargoes themselves. Particular care is needed when washing several tanks that contain dissimilar cargoes, and compatibility should be considered when selecting the destination tank for stripped wash water. The following should be avoided:

- Mixing of slops from Annex I (oil) cargoes with slops from Annex 11 (chemical) cargoes,
- Mixing of slops from incompatible cargoes, and
- Mixing of slops from vegetable oils or fats with chemical slops or petroleum oil slops.

If the ship's cargo tanks are used as slop tanks, care should be taken to avoid introducing slops from cargoes which are incompatible with the tank coating. In this regard, some cargoes which are themselves compatible may, when mixed with water, form acids and thus damage a coating, for example, slops from hydrolytic cargoes in a zinc-coated tank.

Chapter 10

Inert gas systems

GENERAL

In the context of crude and chemical tanker operations and crude and chemical cargoes, an inert gas system may have three distinct uses: preventing fire, preventing chemical reaction and/or maintaining cargo quality. Flammable gases cannot burn in an atmosphere which is deficient in oxygen, which is why inert gas is introduced to the atmosphere to displace air. SOLAS specifies the standards necessary to do this. It may be achieved by using either nitrogen or oil-fired flue gas, with a portable or fixed piping arrangement to supply the inert gas to the cargo tanks, and if applicable, the spaces surrounding the cargo tanks. Mandatory safety requirements for tank atmosphere control are provided in the IBC Code; for example, the system must be able to compensate for normal transportation losses and maintain an overpressure of at least 0.07 bar gauge at all times. There are several types of inert gas systems that can be used on tankers. The most common are as follows:

1. Stored compressed nitrogen,
2. Stored liquid nitrogen,
3. Gaseous nitrogen supplied from shore,
4. Nitrogen generators using pressure swing adsorption (PSA),
5. Nitrogen generators using membrane separation and
6. Oil-fired inert gas generators.

There are occasions when inerting is not appropriate for safety reasons. This is usually when the exclusion of oxygen could create a hazardous reaction with any number of chemicals when shipped in monomer form. Such chemicals (which include acrylic acid, styrene and vinyl acetate) have added inhibitors to prevent polymerisation, during transportation. In order to be effective, however, the inhibitors require the presence of oxygen to be dissolved in the monomer. That oxygen is obtained from the air within the ullage space. Inhibited monomers must therefore be carried in tanks where

DOI: 10.1201/9781003505044-10

the atmosphere has an oxygen level sufficient for the inhibitor to fulfil its purpose.

Requirement of nitrogen used as inert gas on chemical tankers

Most of the nitrogen used as inert gas on tankers is not used for safety reasons but for maintaining cargo quality. Often, shippers will mandate that nitrogen of a particular standard or purity is used for blanketing the cargo once pumped into the vessel's tanks. Small amounts of pure nitrogen may be carried onboard in compressed/liquid form for this purpose; however, in the majority of cases, nitrogen is either pumped onboard from shore-based facilities or is produced using shipboard membrane separators or swing absorption generators. When using an oil-fired inert gas generator, oxygen levels of less than 5% can be obtained, depending on the quality of combustion control and the load on the boiler. The gas must be cooled and scrubbed with water to remove all traces of soot and sulphur acids before being supplied to the cargo tanks. Certain cargoes, however, are chemically reactive and therefore highly sensitive to oxygen concentrations. These can be as low as 2.0% by volume. Alternatively, some cargoes react with the carbon dioxide in flue gases. Other cargoes are overly sensitive to moisture or are liable to discolouration. For these reasons oil-fired flue gas systems are rarely used on chemical carriers when carrying chemical cargoes, as the demand for strict control of atmosphere standards cannot be met. As an indication, the following is a list of potential problems that could occur if oil-fired flue gas systems are used:

- Acid-catalysed hydrolysis (e.g., with esters, acetates or acrylates),
- Acid-catalysed polymerisation (e.g., with allyl chloride),
- Formation of carbonates (e.g., with amines),
- Increased acidity (e.g., contamination of toluene and xylene by carbon dioxide),
- Reaction with water (e.g., with acetone or ethanol) and
- Deactivation of polymerisation inhibitors (e.g., with vinyl acetate).

Pressure swing adsorption (PSA) nitrogen generators

Adsorption is a process in which a substance, usually a gas, accumulates on the surface of a solid to form a very thin film. PSA plants work on the principle that the major constituents of air – nitrogen and oxygen – are adsorbed to a different extent when passed over a carbon molecular sieve material. The amount of each gas adsorbed depends on the duration of exposure. If the system is adjusted correctly, the sieve adsorbs most of the oxygen in the air, allowing the nitrogen to pass through and be collected. The oxygen can then be desorbed (returned to a gas) and exhausted to atmosphere, thereby regenerating the sieve. To provide a continuous nitrogen flow, PSA plants are fitted with two or more interconnected pressurised vessels (called beds) which contain the molecular sieve material. Air is compressed by an oil-free compressor and passed over one set of beds that are adsorbing while the other set of beds is desorbing. During the production cycle, the plant will vent an oxygen-rich waste, which must be exhausted to a safe area. In addition to nitrogen and oxygen, the carbon molecular sieve material also adsorbs a number of other gases, among them carbon dioxide and water vapour. In normal circumstances the carbon dioxide content in air is small, so the presence of carbon dioxide has a negligible effect on the plant operation and any carbon dioxide adsorbed is ejected with the waste gases during the desorption cycle. A number of proprietary sieve materials are water sensitive, and the compressed air must be passed through a dryer to remove atmospheric humidity before passing over the beds. In marine service, the air inlet to the PSA beds must be protected from sea spray. The gas produced by the PSA process can have an oxygen content that varies between 0.1% and 2% by volume depending on the flow rate. Typical plants produce gas with a dewpoint lower than –50°C (–58°F) and a carbon dioxide content of less than 2 ppm by volume.

Membrane separation nitrogen generators

Membrane units work on the premise fact that different gases permeate at different rates through the walls of a thin, hollow membrane. The "slow" gases include methane, nitrogen and carbon monoxide, whereas the "medium" gases are argon and oxygen, and the "fast" gases are water vapour, hydrogen and carbon dioxide. The fact that the two main components of air – nitrogen and oxygen – have different permeation rates means they can be separated. Moreover, because water vapour permeates quickly, the nitrogen produced is also very dry. The membrane unit is made up from bundles of thin hollow fibres which provide a large wall area for separation. The membrane bundles are enclosed in pressure vessel pipes of between 100 and 200 millimetres in diameter; several of these bundles may be arranged in parallel. Clean compressed air is passed through these bundles where the oxygen and water molecules are removed. As the membranes are heat-sensitive, it may be necessary to cool the compressed air before it enters the bundles. The efficiency of the rate of separation depends on the flow rate through the membranes; a control valve is used to regulate the flow and thereby the oxygen content. The flow is adjusted to provide the correct purity of nitrogen required; typically, with an oxygen content variable of between 0.1% and 2% by volume, with water and carbon dioxide contents below 5 ppm. The oxygen-enriched air is then vented as a waste gas, which must be exhausted to a safe area.

Oil-fired inert gas generators

Oil-fired inert gas is acceptable for use with crude and petroleum products, but it has been found that the quality of the inert gas generated by this type of system is not suitable for use with many chemical products as it can adversely affect the cargo quality. It is therefore recommended that when inerting or padding is required by the IBC Code for a particular cargo, nitrogen is used to inert or pad the cargo unless the shipper or supplier has specifically mandated that oil-fired inert gas be used. The basic operating principle of oil-fired plants is that the oxygen content of the air is converted to carbon dioxide through the combustion of oil whilst the nitrogen content remains unchanged. The oil fuel is burnt in a combustion chamber and the combustion (or flue) gas is passed through a water tower (or scrubber) to cool it. The scrubber removes most of the sulphur dioxide, particulates and impurities. This requires contact between the flue gas and copious quantities of seawater.

The gas is then dried by passing through a cooler or alumina bed dryer (or both). Chemical tankers are usually fitted with two non-return valves in series as an equivalent to the deck water seal, thereby avoiding the risk of water carrying over into the cargo. As a further safeguard against backflow, there is usually an isolating valve or a spool piece at each branch connection. The inert gas produced by oil-fired generators depends on the quality of the fuel oil and the efficiency of the combustion and scrubbing processes. These factors influence the amount of sulphides in the inert gas produced, which is why the sulphur content of the fuel is limited in the plant specification. Likewise, inefficient combustion can cause the production of soot, which can clog the scrubber and, more importantly, the dryer system, thereby producing wet and dirty inert gas. If the plant is efficiently burning good quality fuel, the inert gas can be expected to have the following composition.

Carbon dioxide	15%
Oxygen	1.0%
Carbon monoxide	0.1%
Oxides of nitrogen	120 ppm
Hydrogen	100 ppm
Sulphur dioxide and sulphur trioxide	120 ppm
Nitrogen	Balance
Dewpoint	–25°C (–13°F)

NITROGEN PURGING/PADDING AND INERT GAS SYSTEMS

It is the responsibility of the vessel's owner/management to have procedures and checklists in place to control the purging and padding of cargo tanks and enclosed spaces. This is because nitrogen is a "silent killer" and, without proper precautions and training, can (and has) lead (led) to many unnecessary deaths.

Reasons for purging/padding

Purging operations for volatile cargo are often carried out to avoid a flammable atmosphere during loading and to avoid the build-up of static electricity. The purging operation for sensitive cargoes is carried out to avoid any damage to the cargo by moisture or oxygen. In these cases, the tanks are purged down to an oxygen content of 1,000 ppm or less (for some cargoes this may be as much as 50 ppm).

Purging

Some chemical cargoes are so sensitive they require pre-purging in order to avoid any damage to the cargo. In these instances, the maximum oxygen limit can be anywhere between 50 ppm and a maximum of 1,000 ppm. The time used for purging is dependent on the nitrogen supply from onshore. If more than one tank is to be purged at any given time, it is recommended to follow the cascade purging system whenever possible.

Bubbling of nitrogen into the cargo

For some cargos bubbling is used to increase the nitrogen content in the cargo tanks. This is used during the loading of propylene oxide and is done by inserting pure nitrogen into the loading line. The nitrogen is then mixed directly with the cargo.

GUIDANCE FOR VESSELS RECEIVING GASEOUS NITROGEN FROM ONSHORE

It is a frequent practice at chemical loading terminals to control the atmosphere in cargo tanks with nitrogen supplied from onshore. The nitrogen can be for the purpose of drying the ship's tanks and its associated piping system, purging a tank before loading, or padding cargo in a tank. Compressed nitrogen may also be used to propel a line scraper for clearing shorelines into the ship after loading, or for pressing small parcels of cargo out of their shore containers (often railway wagons) and into the ship. During any of these operations, there is potential possibility for an abrupt increase in loading rate of liquid at a few hundred cubic metres per hour to gas at several thousand cubic metres per hour. Agreement on the procedure for handling the nitrogen is paramount and should be part of the preloading checklist between the ship and shore, with emphasis on a clear understanding of the transfer rate and pressure. Although the operation is an important stage in cargo handling, it is also one of the most hazardous as high-pressure gas is being introduced into a tank that is not designed to withstand internal pressure, and whose structure may fail at less than 0.5 bar overpressure. The associated risks of the operation should be thoroughly understood by all members of the ship's staff and the shore loading party. Procedures must be in place to ensure maximum safety during the operation, and all personnel involved should be made conversant with those procedures.

It is possible to over pressurise and even rupture a cargo tank if the nitrogen supply from shore is at too high a flow rate or too great a pressure. Accordingly, there have been incidents where structural damage has occurred. When a liquid is being loaded through the cargo manifold and pipeline system on a chemical tanker, the existing atmosphere in the tank can escape through a vent system that is notably smaller than the liquid filling line. This is because friction and turbulence are far greater impediments to liquid flow than to gas flow. Ships are designed with this in mind. However, when a gas is being introduced through the liquid filling line, especially a gas under pressure that will expand within the tank, the same condition does not apply, and the disparate sizes between inlet and outlet can allow an overpressure to develop. To avoid such an eventuality, the outlet for the existing atmosphere in the tank should be a comparable size or larger than the pipeline supplying the gas. This is usually achieved by having the cargo tank lid or a tank washing hatch open. When vapour control and emission regulations require a closed operation (with the existing tank atmosphere forced to exhaust to shore), the incoming flow of nitrogen must be restricted to a rate equal to or less than the maximum flow of vapour possible through the venting system. If the capacity of the vapour return system is exceeded by the flow of nitrogen into a closed cargo tank, then the only other outlet is through the relief valve which will prevent over pressurisation (though contravening the vapour control regulations). However, if the capacity of both outlets is exceeded, then overpressure will occur and damage to the tank structure may follow.

The pressure or the flow rate of the incoming nitrogen must therefore be controlled. Use of a small hose or a reducer prior to the manifold will limit the flow rate, but pressure must be controlled by the shore. The ship's manifold valve is designed to control liquid flow. However, in an emergency the manifold valve can be used as a rapid safety stop; pressure surge in a gas is not as violent as in a liquid and the nitrogen supply hoses are designed to sustain this pressure.

Quantity of nitrogen required

The anticipated quantity of nitrogen required for any particular purpose must be carefully evaluated, taking into consideration the number of tanks that require maintenance of nitrogen blanket; the anticipated duration of the voyage for which the nitrogen blanket has to be maintained; the production and storage capacity of the nitrogen generator (where installed); any interim ports in which stocking up may be possible; and the expected weather conditions during the voyage.

Drying or purging an empty tank that has been cleaned and gas freed

During the pretransfer planning conference, the volume of nitrogen required should always be calculated and agreed (tank volume multiplied by number of volumes to reach the desired level of dryness or oxygen exclusion), and the flow rate agreed. Table 10.1 shows the volume of nitrogen that can be received in 1 minute through a known size of pipe at a known pressure. The second figure in brackets indicates the associated hourly rate which should be mentally compared to a liquid loading rate. Note that these tables are intended to be indicative only, and any discrepancies are due to the rounding of figures.

If padding has to be performed after loading, planning and good communication are essential. The supply rate must not exceed the vent capacity of the cargo tank. The vapour space in a loaded tank is small, so over pressurisation can occur very suddenly, especially if cargo is forced into the vent lines which then become restricted or blocked and add to the rapid increase in tank pressure. A safe practice is to introduce the nitrogen directly into the ullage space via a small diameter connection at the top of the tank (PV mast) or into the ship's cargo line, preferably using the ship's own equipment and gas supply at low pressure, or if necessary, from shore through a small diameter connection to restrict the flow. Introducing it from shore through the manifold valve and the tank filling line after clearing the shore loading line presents a higher risk and provides

Table 10.1 Cubic metres of gas at various gauge pressures (received in 1 minute (and 1 hour) through hoses of various diameters)

	200 mm (8")	*150 mm (6")*	*100 mm (4")*	*50 mm (2")*	*25 mm (1")*
5.2 bar (75 psi)	1771 (106,000)	914 (55,000)	343 (20,600)	67 (4,000)	12 (740)
3.4 bar (50 psi)	1,286 (77,000)	662 (39,700)	243 (14,600)	48 (2,900)	9 (530)
2.1 bar (30 psi)	886 (53,000)	171 (10,300)	171 (10,300)	33 (2,000)	6 (360)
0.7 bar (10 psi)	471 (28,300)	214 (12,900)	80 (4,800)	16 (1,000)	3 (170)

extraordinarily little time to react. Pressure gauges and sensors should be closely monitored during the operation, and the ship's officer should be in direct supervision throughout. The operation should be stopped when a slight overpressure exists in the ullage space, but before reaching the tank pressure relief valve setting.

Using nitrogen to press a product out of shore tanks into the ship, or for clearing the shoreline into the ship after loading

These operations may also be undertaken using other compressed gas, usually air, but the process and the inherent risk of over pressurisation are the same. The gas pressure used for these an operation varies but can range between 2.5 and 5 bar. During a line clearing operation, it is important that terminal staff react promptly when the scraper is caught in its trap, in order to avoid all the compressed propelling gas entering a loaded cargo tank. The point of greatest concern is when the supply into the ships tank changes from liquid to compressed gas, and the tank filling rate increases dramatically. It will be seen from Table 10.1 that a significant volume of gas will be received in a few seconds through the large liquid filling line. Over pressurisation of a closed tank can occur in seconds, especially when the distance from the manifold to the tank is small or the vapour space in the tank is limited.

Discharge of cargo

When discharging cargoes that have to be carried under a blanket of nitrogen, it may be necessary to ensure that no air is drawn into the tank. Therefore, an overpressure of nitrogen should be maintained as the liquid level falls, using stored compressed gas or from a nitrogen generator on board, and be introduced into the tank ullage space. If it is necessary to obtain nitrogen from the shore, it is essential that the pretransfer discussion includes agreement on the nitrogen flow rate and pressure to be used. Although the overpressure required is no more than about 0.2 bar, it is usual for the shore nitrogen supply system to be well above this figure, as high as 7 bars. Particularly in the early stages when the ullage space is still small, it is possible for the flow rate to exceed the tank venting capacity, and for an overpressure to develop. A safe procedure is to use a pressure-reducing device on the nitrogen supply pipeline, and to have a calibrated gauge showing the pressure in the pipeline. There should be constant communication with the terminal, and the ship should monitor cargo tank ullage space pressure throughout. The ship's staff and shore personnel should agree in writing on the inert gas supply, specifying the volume required and the flow rate in cubic metres per minute. The sequence of operating valves before beginning the operation and after completion should be agreed, so that the ship remains in control of the flow. Attention should be given to the adequacy of open vents on a tank to avoid the possibility of over pressurisation. The tank pressure should be closely monitored throughout the operation. The ships agreement should be sought when the terminal wishes to use compressed nitrogen or air as a propellant, either for a line scraper to clear shore pipelines into the ship or to press cargo out of shore containment. The ship should be informed of the pressure to be used and the possibility of receiving gas into the cargo tank.

PADDING OF LOADED TANKS

After the tank has been loaded, it might be necessary to add additional nitrogen to create an effective blanket over the cargo. Similar to the purging operation, the padding of tanks may create risks associated with tank rupture and cargo overflow. The vessel must therefore be in absolute control of the entire padding operation and the duty officer should personally supervise the operation of the manifold valve and must control the start as well as the flow of nitrogen. It should be borne in mind that throttling of conventional vessel manifold valves is an ineffective way of controlling the flow or restricting the gas volume. The tank pressure is to be closely monitored. To control the pressure into a loaded cargo tank and especially in situation where the tank is filled up to 95%/98%, the use of only stripping line open for filling of nitrogen is an easy and safe way of controlling the flow into the tank. This method can be used on all submerged type cargo pumps.

Onboard nitrogen generator safety check (for vessels fitted with a nitrogen generator)

It is expected that the nitrogen generator onboard be kept in good working condition at all times. Normally generators fitted onboard vessels operate in either a 99.9% or 95% mode. This means it can be used for padding highly reactive chemical cargos as well as serving as a substitute for conventional inert gas systems for loading and discharging any Annex I type cargoes. The calibration of some analysers can be complicated; therefore, it is imperative that the ship's staff responsible for overseeing and maintaining the nitrogen generator are fully familiarised with the system upon joining the vessel. Regular operation of the system will usually detect any malfunctioning at an early stage. To keep the system in a state of readiness, it should be tested every month in both 99.9% and 95% mode. In accordance with the onboard standard operating procedure (SOP), the stripping line should be opened for purging and padding, and the valve put online with the dropline closed. The oxygen analyser should be calibrated regularly to ensure it provides accurate readings; the sample line should also be clean and free of moisture.

The tests should not be limited to only running the system. The nitrogen produced should be delivered to deck. This can be done by opening any flange on the nitrogen branch line on deck. This will ensure smooth operation of the pneumatically operated deck valves as well as the pressure control valve and the main supply valve located in the engine room. This practice will minimise the opportunity for the valves to seize due to long periods of non-use. It will also put the entire electronic control system under test. In addition, manual operation of the valves should be performed so that the valves can be operated manually if required. All relevant maintenance related to the compressors, filters and auto drain valves must be carried out in accordance with the ship's SMS. When confirming the functioning of the nitrogen generator, always refer to the manufacturer's manual. In most cases, the initial step is to check the alarm oil level, drain valve and belt, to identify any indications of oil leakage during the operation of the compressor. Second, in accordance with the manufacturer's guidance, change the oil of the compressor as per the running hour schedule or at least once every 12 months. Check the operation of the drain valve and overhaul as required. Third, check the filter differential is maintained within acceptable limits. When the filter differential exceeds 0.07 MPa, change the element. The fourth step is to calibrate the oxygen analyser before every operation. Pass instrument air through the analyser to check alarm functioning during operation. The fifth and final step is to check the operation of the system's alarms. This should be done at least once every three months by simulating the alarm condition. Additional steps to be taken depending on the age and condition of the generator might include removing the heater from its shell once each year and/or replacing the charcoal bed once every five years.

INERTING CRUDE TANKS BEFORE LOADING

Prior to arrival at the oil loading port, all empty tanks are to be inerted and a check made with a portable oxygen analyser to ensure the tanks do not contain more than 8% oxygen by volume. Some terminals may limit this to 5% oxygen by volume. The mast riser cowl wire mesh should be inspected, and cleaned, if necessary, preferably before each voyage.

Loading

During the loading of inert gas, the plant must be shut down and the deck main isolating valve closed. The inert gas pressure recorder is to be kept onboard to record any abnormal pressure surges during the loading procedure.

On passage

It is important to maintain a slight positive pressure in the ullage spaces of cargo tanks to avoid the ingress of air through the P/V valves. It may be necessary to operate the inert gas plant periodically for short periods to maintain this condition. The inert gas pressure should be logged in the deck log at the end of each watch.

Discharge

The inert gas plant must be started prior to the commencement of discharge and **operated continuously throughout the discharge operation.** To be effective, the main blower should have a capacity at least 25% in excess of the total pumping capacity to enable positive pressure to be maintained in the tanks. Both the oxygen content and pressure of the inert gas must be continuously recorded during discharge. A separate risk assessment is to be done should the vessel be required to ballast any cargo tank in port. The Chief Officer must also be informed whenever the shore loading party requires the shoreline to be flushed with seawater that is delivered by the ship's cargo pump.

Tank cleaning

Prior to the commencement of tank cleaning with water, the atmosphere in each tank must be checked with a portable oxygen analyser and the oxygen content established as being at least 8% by volume or less throughout. Testing should be conducted at three levels in each tank and, where possible, from more than one sampling point. Inert gas must be supplied to the tanks throughout washing.

GAS-FREEING

If any tanks are required to be gas freed, they must first be purged with inert gas to reduce the hydrocarbon gas content to a maximum of 2% by volume as determined by a properly calibrated Tankscope. The purpose is to reduce the hydrocarbon gas content to below the critical dilution line so that subsequent introductions of air will not result in a flammable mixture developing. Tanks which are undergoing gas-freeing must be isolated from the rest of the inert gas system.

Inert gas safety controls

The inert gas control panel provides constant information regarding system pressures together with high- and low-pressure alarms. It must not be switched off at any time when the ship has petroleum cargo or slops onboard. The deck water seal is the ultimate safety barrier between the cargo system and the engine room, and it is essential that the water is kept at the correct level at all times.

Safety checks when the inert gas plant is shut down

Whenever the inert gas plant is shut down, i.e., during loading, and on passage, the vent valve located between the deck water seal and the inert gas blowers must be open. The normal level of the deck seal must be prominently marked. The following safety checks should be carried out at sea whenever the inert gas plant is not operational. The water supply and water level in the deck seal should be ascertained at regular intervals, and at least once per day depending on weather conditions. In cold weather, always ensure the arrangement for preventing the freezing of sealing water in the deck seal, pressure/vacuum breaker etc., are in good working order. Before the pressure in the inerted cargo tanks drops to 100 mm, the tanks should be repressurised with inert gas.

System test schedule

Prior to arrival at the port of discharge, the system testing programme should be carried out. This will vary from one vessel to the next, but in general, the programme should include the following:

- All alarms and trips,
- The functioning of the flue gas isolating valves,
- The operation of all remote or automatically controlled valves,
- Function test of deck water seals and non-return valves,
- The vibration level of the inert gas system blowers,
- Checks for gas leaks in the inert gas system both before and after the deck seal,
- The interlocking capability of soot blowers,
- Checks for any gas leaking from the tanks (e.g., Butterworth covers, and ullage ports),
- Operation of the P/V valve and
- The calibration and operation of the fixed oxygen analyser.

Some company SOPs require the fixed oxygen analyser to be calibrated immediately prior to the use of the inert gas system. Where this is the case, the vessel must comply with this requirement.

Maintenance programme

Always follow the manufacturer's guidelines in relation to preventative maintenance. The maintenance programme forms an integral part of the ship's SMS.

Exchange of tank atmosphere

There are three principal arrangements.

Arrangement	*Inlet point*	*Outlet point*	*Principle*
1	Top	Top	Dilution
2	Bottom	Top	Dilution
3	Top	Bottom	Displacement or dilution

The main difference between dilution and displacement is that with the dilution method the velocity of the incoming gas should be high while in displacement the velocity of the incoming gas is low. With the latter, the velocity of the incoming gas should be as low as 250 mm water gauge. This is also the more efficient of the two methods. In the displacement method it is important that the bellmouth is free of any liquid, which will create a seal and obstruct the passage of inert gas. Therefore, the tanks have to be well stripped. When inerting the cargo tanks with fresh air, it is important to remember the oxygen content must be below 8% at all times.

Composition and quality of inert gas

SOLAS requires that inert gas systems are capable of delivering inert gas with an oxygen content in the inert gas main of not more than 5% by volume at any required rate of flow; and of maintaining a positive pressure in the cargo tanks at all times with an atmosphere having an oxygen content of not more than 8% by volume except when it is necessary for the tank to be gas free. In certain ports the maximum oxygen content of inert gas in the cargo tanks may be set at 5% to meet particular safety requirements, such as the operation of a vapour emission control system. When such a limitation is in place, the ship should discuss these requirements in the prearrival information exchange. Efficient scrubbing of the inert gas is essential, particularly for the reduction of the sulphur dioxide content. High levels of sulphur dioxide increase the acidic characteristic of the inert gas, which is harmful for personnel and may cause accelerated corrosion to the structure of the ship.

Inert gas system failure

SOLAS requires that each ship fitted with an inert gas system be provided with detailed instruction manuals covering its operation, safety and maintenance requirements, and the occupational health hazards relevant to the installed system and its application to the cargo tank system. The manual must include guidance on procedures to be followed in the event of a fault or failure of the inert gas system.

Action to be taken on failure of the inert gas system

In the event that the inert gas system fails to deliver the required quality and quantity of inert gas, or to maintain a positive pressure in the cargo tanks and slop tanks, action must be taken immediately to prevent any air being drawn into the tanks. All cargo and/or ballast discharge from inerted tanks must be stopped. The inert gas isolating valve closed, the vent valve between it and the gas pressure regulating valve (if provided) opened and immediate action taken to repair the inert gas system. Masters are reminded that national and local regulations may require the failure of an inert gas system to be reported to the Port State authority, terminal operator, and to Flag State Administrations. In the event that the inert gas system is unable to meet the operational requirements set by SOLAS, and assessments indicate that it is impracticable to effect a repair, then cargo discharge, deballasting and necessary tank cleaning can only resume once the emergency conditions laid down in the "IMO Guidelines on Inert Gas Systems" are complied with. In summary, the guidelines mandate that

1. In the case of tankers engaged in the carriage of crude oil, it is essential that the tanks be maintained in the inerted condition to avoid the danger of pyrophoric iron sulphide ignition. If it is assessed that the tanks cannot be maintained in an inerted condition before the inert gas system can be repaired, an external supply of inert gas should be connected to the system to avoid air being drawn into the cargo tanks, and/or
2. In the case of the carriage of products, if it is considered impracticable to effect repair of the inert gas system, cargo discharge may only be resumed if an external supply of inert gas is connected, or the following precautions are taken:
    a. That approved device, or flame screens, to prevent the passage of flame into cargo tanks are fitted and checked to ensure that they are in good order,
    b. The valves on the mast risers are opened,
    c. No free fall of water or slops is permitted and
    d. No dipping, ullaging, sampling or other equipment should be introduced into the tank until a period of 5 hours since injection of inert gas ceased. If essential for the safety of the operation, this should be done only after 30 minutes have elapsed and all metal components should be securely earthed.

Chapter 11

Tanker hazards and safety

The word "health" brings many things to mind. Maintaining good health involves a nutritious and balanced diet, regular exercise, vaccinations against diseases and visiting a physician whenever things are not as they should be. Our health is indicative of how well our body is functioning and is a measure of our general quality of life. We can also appreciate health in a broader sense. Human and environmental health involves understanding the impact of environmental and human-made hazards and establishing and maintaining ways of protecting human health and ecological systems against these same hazards. This includes studying the impact of human-made chemicals on marine life as well as on human health. It involves understanding how the environment influences the spread of disease and infection. Ships are recognised as one of the main ways that diseases and infections can spread around the world. Vectors such as rats and insects can find their way onboard unnoticed in the ship's holds and cargoes. Inevitably, modern-day seafarers face countless environmental hazards every day. To better understand them, we can think of them as falling into any one of four categories: physical, chemical, biological and cultural.

PHYSICAL, CHEMICAL, BIOLOGICAL AND CULTURAL HAZARDS

Physical hazards are physical processes that occur naturally in the environment. These include natural disaster events such as earthquakes, tornadoes, volcanoes, blizzards, landslides and droughts. Not all physical hazards are discrete events; some are ongoing, such as ultraviolet (UV) radiation. UV radiation is considered a hazard because it damages DNA and can cause human health issues including skin cancer and cataracts.

Chemical hazards can include both natural chemicals found in the environment and those that are human made. Human-made chemical hazards include many of the synthetic chemicals produced by

DOI: 10.1201/9781003505044-11

industry such as disinfectants, pesticides and plastics. Some chemical hazards occur naturally in the environment, like the heavy metals lead and mercury. Some organisms even produce natural chemicals that are an environmental hazard, such as the compounds in peanuts and dairy that cause allergic reactions in humans.

Biological hazards are derived from ecological interactions between organisms. Viruses, bacterial infections, malaria and tuberculosis are all examples of biological hazards. When these pathogens and diseases are transferred between organisms, it is defined as an infectious disease. Humans and animals suffer from these diseases and pathogens because we become parasitised by another organism. Whilst this is clearly injurious to health, it is an entirely natural process.

Cultural hazards, also known as social hazards, result from our location, socioeconomic status, occupation and behavioural choices. For instance, smoking cigarettes is scientifically proven to be hazardous to health. Even so, many people *choose* to smoke as this is a behavioural choice. Alternatively, living in an area with high rates of crime is a hazard based on our location. Similarly, diet, exercise habits and primary mode of transportation all influence our health and the health of the environment around us.

There are some hazards which we can avoid but there are many others which we simply cannot. It is our responsibility to ensure that we eliminate – as far as is practicable – those hazards which we can control and reduce our exposure to those hazards which are beyond our control. There are various processes and procedures that can be used to both eliminate and reduce the risk of hazards in the workplace. Living and working onboard any ship is inherently dangerous. The marine environment is a very unforgiving place. When pushed to her limits, Mother Nature has the power to destroy and sink even the largest and best designed vessels.

CHEMICAL HAZARDS

By their very purpose, oil and chemical tankers are by far the two most dangerous types of ships in operation. Chemical reactivity hazards exist even when there is no intention to perform chemical reactions. Indeed, chemical reactivity hazards are manifest through materials which become chemically unstable for a variety of reasons; or when intended chemical reactions get out of control; and in the worst case when unintended chemical reactions take place due to the accidental mixing of chemicals that are normally kept separate.

CORROSION HAZARDS

Alternatively, corrosion hazards exist when handling or transporting corrosives such as acids, alkalis (bases or caustics) and halogens. A corrosive material is a reactive compound or solution that produces a destructive chemical change in the material upon which it is acting. On contact, a corrosive material may destroy metals, body tissues, plastics and other materials. Those corrosive solutions that have the greatest concentration of hydrogen ions (H) are the strongest acids, whilst those that have the greatest concentration of hydroxyl ions (OH) are the strongest bases. The measure of acidity and basicity is pH: strong acids have a low pH (with a maximum of 1), whilst strong bases have a high pH (maximum of 14). Neutrality is given the pH of 7. Examples of common corrosives include acids such as hydrochloric acid, nitric acid, sulphuric acid; bases, which include potassium hydroxide, sodium hydroxide (caustic soda); halogens, such as chlorine, bromine.

FLAMMABILITY HAZARDS

The danger of explosion exists when flammable materials are mixed with air, where they form an explosive mixture. This can occur during storage, movement, process, production and the manufacture of flammable materials. The primary safety requirement is for the operator to prevent conditions where a flammable mixture is released to the atmosphere. However, as there is always some element of risk that such a situation may take place, special measures must be taken in respect of electrical and non-electrical apparatus, to prevent the possible ignition of flammable or explosive atmospheres. The employment of these measures should safeguard both the plant (or installation) and, more importantly, human life, as ignition can only occur when both a flammable atmosphere and the means for an ignition exist simultaneously. Such ignition may occur following an arc, spark or hot surface, such as during the use of electrically powered equipment. It should also be recognised that non-electrical equipment may also form the source of ignition. In addition, ignition could also be initiated by frictional sparking and electrostatic action. Arcs can result from the discharge of stored energy or from switching contacts. Hot surfaces sufficient to cause ignition can arise from electrical enclosures or components. Other sources of ignition energy include open flames (including matches and lighters), stray electric currents, lightning, compression, engine exhausts, heat from chemical reactions, spontaneous combustion, friction or heat from mechanical equipment and heat from the sun.

There are three components which are necessary for ignition to occur. These are a flammable material, liquid or gas, oxygen (in air) and a source of ignition. Technically, a fourth factor is also needed: a chemical reaction.

Historically, the first three components were referred to as the "fire triangle." As the understanding of the science of fire has developed, the fourth component has become increasingly prominent, leading to the term, "fire tetrahedron." Whichever model is used, it is important to understand that without the presence of all three (or four) components, in exactly the right quantity, it is chemically impossible for ignition to occur. This means the underlying principle of fire prevention – and fire extinction – is the removal of one or more of these components. By removing the flammable source, we are starving the fire of material to burn; removing oxygen (air) will suffocate the fire. By not introducing a source of ignition, there is no way a fire could ignite in the first place.

ELECTROSTATIC CHARGE HAZARDS

It is worth noting that on oil and chemical tankers, one of the main sources of ignition is static electricity. Electrostatic charge is often developed due to the relative motion between two dissimilar substances. This may dissipate to earth or less charged objects resulting in a spark. In the presence of flammable gases or hydrocarbon vapours, this can cause fire, explosion, detonation and loss of plant, property and personnel. Technically, static charge is generated from the movement of electrons when two dissimilar substances come into close contact and then separate. The transfer of electrons occurs across the interface between the two surfaces. This contact leads to line-ups of opposite electrical charges on two adjacent substances, at the boundary between them. Magnitude and polarity of charge being function of area between the contact surfaces, the type of materials, surface temperature and separation velocity. A separation of two layers with opposite charges leads to potential difference. Electrical conductivity of material determines the time required for the charge to become neutralised and the separation rate determines the time available for discharge. The atmosphere may become ionised if the field strength exceeds critical level, which may lead to a discharge to proximal earthed or less charged objects as a complete spark. This occurrence is responsible for the majority of industrial fires and explosions. Alternatively, a partial breakdown – or corona discharge – may occur depending on the configuration of the two surfaces which act as a capacitor and field strength.

TOXICITY HAZARDS

Any number of commonly shipped toxic substances can cause profoundly serious health effects in an exposed individual. The degree of hazard associated with any toxic substance is related to the specific substance (whether powder, gas/vapour, liquid or solid) the casualty is exposed to, as well as the concentration of the substance, the route into the body and the amount absorbed by the body (i.e., the dose). Individual susceptibility also plays a key role. The health impact may occur immediately, or the effects may be delayed. Health effects that occur immediately after a single exposure are called "acute" effects. In other cases, health effects may not manifest until sometime has passed following after the exposure. This is called a "chronic" effect. A chronic effect may occur hours, days, months or even years after exposure. Acute effects are caused by a single, high exposure. Chronic effects tend to occur over a longer period of time and involve lower exposures (e.g., exposure to a smaller amount over time). Some toxic substances can have both acute and chronic health effects. To avoid exposure, personnel must be fully trained and competent (i.e., suitably qualified and experienced person [SQEP]) to operate in environments where toxic substances are handled. Furthermore, personnel who may be directly affected by toxic substances must don the mandatory personal protective equipment outlined in the ship's/terminal's standard operating procedure (SOP). This typically includes wearing outer hazmat garments (splash apron), gloves, eye goggles or facemask, and protective shoes or boots. Specific requirements will depend on the substance, the working environment, location and any other relevant factors which may cause injury or hazards to health.

VAPOUR LEAKS AND CLOUDS

When loaded into the cargo tanks, the pressure of the vapour phase is maintained at a constant level, usually slightly above atmospheric pressure. The external heat that passes through the tank insulation generates convection currents within the cargo. This can lead to heated gases rising to the surface of the tank where it vaporises. The heat necessary for vaporisation comes from the cargo itself. For example, when loading liquefied natural gas (LNG), as long as the vapour is continuously removed by maintaining the pressure at a constant, the LNG will remain at its boiling temperature. If the vapour pressure is reduced by removing more vapour that is generated, the LNG temperature will decrease. In order to make up the equilibrium pressure corresponding to its temperature, the vaporisation of LNG is accelerated, resulting in an increased heat transfer of LNG to vapour. Cargo vapour, whether toxic or flammable, should be vented to the atmosphere with extreme caution, taking account of local regulations and the prevailing weather conditions. If the temperature of the vented vapour is below the atmospheric dew point, clouds of condensed water vapour will form. Condensed water vapour (fog) is heavier than air. The cargo vapour may or may not be heavier than air, depending on the atmospheric temperature. The cargo vapour cloud is likely to be oxygen deficient and should only be entered into by personnel wearing self-contained breathing apparatus (SCBA). Furthermore, it should never be assumed that the cargo vapour is contained entirely within the boundaries of the visible water vapour cloud. If the cargo vapour is heavier than air, it may accumulate on deck and enter accommodation spaces. Standard precautions should therefore be observed. In some cases, it may be possible to heat the vapour prior to venting to reduce its density and to assist dispersion. If such facilities are provided, they should be used, although in some cases, venting may be prohibited.

VOID SPACES/DUCT KEELS AND PIPE TUNNELS

Because of restricted natural ventilation, these spaces are more likely than not to be oxygen deficient. In addition, where they are adjacent to cargo holds and ballast tanks, hydrocarbon vapours and inert gases may leak into them. It must be recognised that the rescue of an unconscious or injured casualty in these locations may be exceedingly difficult. Hydrocarbon vapours may also be released from adjacent cargo tanks through leaks in pipelines or hairline cracks in the tank structure. It is therefore essential that regular checks are made of these spaces for the presence of hydrocarbons. Some ships will be equipped with automatic detectors and recording devices for this purpose. Even so, manual checks and inspections should be made for increased peace of mind. Ships that do not have such equipment must carry out manual checks at least weekly and the results recorded in the table within the deck logbook. Where the vessel's classification society rules do not require permanent lighting systems in these spaces to be isolated during gas trading, the following procedures are recommended in order to eliminate the hazards that could result from damaged flame proof fittings. The space should be tested and proven gas free before the lights are switched on. The mechanical ventilation system should be in operation before the lights are switched on and must remain in operation until after the lights are switched off. Fortunately, to limit the hazards associated with vapours in cargo tanks, there are four main types of tank hazard controls: inerting, water padding, use of drying agents and monitoring techniques.

MATERIAL SAFETY DATA SHEETS (MSDS)

Material Safety Data Sheets or MSDS are an especially important and useful source of information for crew members and operational personnel involved in handling chemicals and cargo products. The MSDS should be supplied directly by the manufacturer, supplier or shipper for each type of cargo to be loaded onboard the vessel. In the event the MSDS are unavailable, the ship must be notified immediately for further instructions prior to any cargo being loaded. A letter of protest (LoP) may also be issued. The MSDS for the products being carried must be displayed prominently onboard the vessel and all crew briefed in order that all crew members are familiar with the properties and handling of the cargoes onboard. The data sheets comprise different sections and the layout may differ slightly from supplier to supplier; however, the information contained within them is the same and can be summarised as follows:

1. Identification of substance (including the trade name and manufacturer),
2. Composition and formula,
3. Principal hazards (e.g., toxic, and carcinogenic),
4. First aid measures,
5. Firefighting measures,
6. Accidental release measures,
7. Handling and storage,
8. Exposure controls and personal protective equipment (PPE and threshold limit values),
9. Physical and chemical properties,
10. Stability and reactivity,
11. Toxicological information,
12. Ecological information,
13. Disposal considerations,
14. Transport information (e.g., UN numbers) and
15. Further information.

It is critically important that the MSDS is specific to the actual cargo product being loaded, and not a generic one, in order to establish all the cargo's components. Once received, it must be read and fully understood as the health and safety of those working with the cargo and others is dependent on the safe handling and suitable precautions of the chemical in question (refer to Table 11.1).

CARGO CONTROL AND MEASUREMENT INSTRUMENTS ONBOARD CHEMICAL TANKERS

In order to maintain a proper control of the tank atmosphere and to check the effectiveness of gas-freeing, especially prior to tank entry, several different gas measuring instruments need to be available for use. Which one to use will depend on the type of atmosphere to be measured. Tank atmosphere sampling lines should be in all respects suitable for, and impervious to, the gases present and should be resistant to the effects of hot wash water. Instruments themselves should be used in accordance with the manufacturers' instructions. Much of the monitoring and measurement during cargo operations on chemical tankers remains reliant on the human interpretation of information, and subsequent decisions are made on the basis of training and experience. Those factors will continue to be

Table 11.1 Extract from the UN number database

DANGEROUS GOODS LIST

UN No. (1)	Name and description (2)	Class or division (3)	Subsidiary risk (4)	UN packing group (5)	Special provisions (6)	Limited quantities (7)	Packagings and IBCs		Portable tanks	
							Packing instruction (8)	Special provisions (9)	Portable tank instruction (10)	Portable tank special provisions (11)
0004	AMMONIUM PICRATE dry or wetted with less than 10% water, by mass^	1.1D				NONE	P112(a) (b) or (c)	PP26		
0005	CARTRIDGES FOR WEAPONS with bursting charge^	1.1F				NONE	P130			
0006	CARTRIDGES FOR WEAPONS with bursting charge^	1.1E				NONE	P130 LP101	PP67 L1		
0007	CARTRIDGES FOR WEAPONS with bursting charge^	1.2F				NONE	P130			
0009	AMMUNITION, INCENDIARY with or without burster, expelling charge or propelling charge^	1.2G				NONE	P130 LP101	PP67 L1		
0010	AMMUNITION, INCENDIARY with or without burster, expelling charge or propelling charge^	1.3G				NONE	P130 LP101	PP67 L1		
0012	CARTRIDGES FOR WEAPONS, INERT PROJECTILE or CARTRIDGES, SMALL ARMS^	1.4S				NONE	P130			

(continued)

Table 11.1 (Cont.)

DANGEROUS GOODS LIST

UN No. (1)	Name and description (2)	Class or division (3)	Subsidiary risk (4)	UN packing group (5)	Special provisions (6)	Limited quantities (7)	Packagings and IBCs: Packing instruction (8)	Packagings and IBCs: Special provisions (9)	Portable tanks: Portable tank instruction (10)	Portable tanks: Portable tank special provisions (11)
0014	CARTRIDGES FOR WEAPONS, BLANK or CARTRIDGES, SMALL ARMS, BLANK^	1.4S				NONE	P130			
0015	AMMUNITION, SMOKE with or without burster, expelling charge or propelling charge^	1.2G			204	NONE	P130 LP101	PP67 L1		
0016	AMMUNITION, SMOKE with or without burster, expelling charge or propelling charge^	1.3G			204	NONE	P130 LP101	PP67 L1		
0018	AMMUNITION, TEAR-PRODUCING with burster, expelling charge or propelling charge^	1.2G	6.1 8			NONE	P130 LP101	PP67 L1		
0019	AMMUNITION, TEAR-PRODUCING with burster, expelling charge or propelling charge^	1.3G	6.1 8			NONE	P130 LP101	PP67 L1		
0020	AMMUNITION, TOXIC with burster, expelling charge or propelling charge^	1.2K	6.1		274	NONE	P101			
0021	AMMUNITION, TOXIC with burster, expelling charge or propelling charge^	1.3K	6.1		274	NONE	P101			
0027	BLACK POWDER (GUNPOWDER), granular or as a meal^	1.1D				NONE	P113	PP50		

0028	BLACK POWDER (GUNPOWDER), COMPRESSED or BLACK POWDER (GUNPOWDER), IN PELLETS^	1.1D	NONE	P113	PP51
0029	DETONATORS, NON-ELECTRIC for blasting^	1.1B	NONE	P131	PP68
0030	DETONATORS, ELECTRIC for blasting^	1.1B	NONE	P131	
0033	BOMBS with bursting charge^	1.1F	NONE	P130	
0034	BOMBS with bursting charge^	1.1D	NONE	P130 LP101	PP67 L1
0035	BOMBS with bursting charge^	1.2D	NONE	P130 LP101	PP67 L1
0037	BOMBS, PHOTO-FLASH^	1.1F	NONE	P130	
0038	BOMBS, PHOTO-FLASH^	1.1D	NONE	P130 LP101	PP67 L1
0039	BOMBS, PHOTO-FLASH^	1.2G	NONE	P130 LP101	PP67 L1
0042	BOOSTERS without detonator^	1.1D	NONE	P132 (a) or (b)	
0043	BURSTERS, explosive^	1.1D	NONE	P133	PP69
0044	PRIMERS, CAP TYPE^	1.4S	NONE	P133	
0048	CHARGES, DEMOLITION^	1.1D	NONE	P130 LP101	PP67 L1
0049	CARTRIDGES, FLASH^	1.1G	NONE	P135	
0050	CARTRIDGES, FLASH^	1.3G	NONE	P135	
0054	CARTRIDGES, SIGNAL^	1.3G	NONE	P135	
0055	CASES, CARTRIDGE, EMPTY, WITH PRIMER^	1.4S	NONE	P136	
0056	CHARGES, DEPTH^	1.1D	NONE	P130 LP101	PP67 L1
0059	CHARGES, SHAPED without detonator^	1.1D	NONE	P137	PP70

(*continued*)

Table 11.1 (Cont.)

DANGEROUS GOODS LIST

UN No. (1)	Name and description (2)	Class or division (3)	Subsidiary risk (4)	UN packing group (5)	Special provisions (6)	Limited quantities (7)	Packagings and IBCs: Packing instruction (8)	Packagings and IBCs: Special provisions (9)	Portable tanks: Portable tank instruction (10)	Portable tanks: Portable tank special provisions (11)
0060	CHARGES, SUPPLEMENTARY, EXPLOSIVE^	1.1D				NONE	P132 (a) or (b)			
0065	CORD, DETONATING, flexible^	1.1D				NONE	P139	PP71 PP72		
0066	CORD, IGNITER^	1.4G				NONE	P140			
0070	CUTTERS, CABLE, EXPLOSIVE^	1.4S				NONE	P134 LP102			
0072	CYCLOTRIMETHYLENETRINITRA- MINE (CYCLONITE; HEXOGEN; RDX), WETTED with not less than 15% water, by mass^	1.1D			266	NONE	P112(a)	PP45		
0073	DETONATORS FOR AMMUNITION^	1.1B				NONE	P133			
0074	DIAZODINITROPHENOL, WETTED with not less than 40% water, or mixture of alcohol and water, by mass^	1.1A			266	NONE	P110 (a) or (b)	PP42		
0075	DIETHYLENEGLYCOL DINITRATE, DESENSITIZED with not less than 25% non-volatile, water-insoluble phlegmatizer, by mass^	1.1D			266	NONE	P115	PP53 PP54 PP57 PP58		

0076	DINITROPHENOL, dry or wetted with less than 15% water, by mass^	1.1D	6.1		NONE	P112 (a), (b) or (c)	PP26
0077	DINITROPHENOLATES, alkali metals, dry or wetted with less than 15% water, by mass^	1.3C	6.1		NONE	P114 (a) or (b)	PP26
0078	DINITRORESORCINOL, dry or wetted with less than 15% water, by mass^	1.1D			NONE	P112 (a), (b) or (c)	PP26
0079	HEXANITRODIPHENYLAMINE (DIPICRYLAMINE; HEXYL)^	1.1D			NONE	P112 (b) or (c)	
0081	EXPLOSIVE, BLASTING, TYPE A^	1.1D			NONE	P116	PP63 PP66
0082	EXPLOSIVE, BLASTING, TYPE B^	1.1D			NONE	P116 IBC100	PP61 PP62 PP65 B9
0083	EXPLOSIVE, BLASTING, TYPE C^	1.1D		267	NONE	P116	
0084	EXPLOSIVE, BLASTING, TYPE D^	1.1D			NONE	P116	
0092	FLARES, SURFACE^	1.3G			NONE	P135	
0093	FLARES, AERIAL^	1.3G			NONE	P135	
0094	FLASH POWDER^	1.1G			NONE	P113	PP49
0099	FRACTURING DEVICES, EXPLOSIVE without detonator, for oil wells	1.1D			NONE	P134 LP102	
0101	FUSE, NON-DETONATING^	1.3G			NONE	P140	PP74 PP75
0102	CORD (FUSE), DETONATING, metal clad^	1.2D			NONE	P139	PP71
0103	FUSE, IGNITER, tubular, metal clad^	1.4G			NONE	P140	
0104	CORD (FUSE), DETONATING, MILD EFFECT, metal clad^	1.4D			NONE	P139	PP71
0105	FUSE, SAFETY^	1.4S			NONE	P140	PP73
0106	FUZES, DETONATING^	1.1B			NONE	P141	

(continued)

Table 11.1 (Cont.)

DANGEROUS GOODS LIST

UN No. (1)	Name and description (2)	Class or division (3)	Subsidiary risk (4)	UN packing group (5)	Special provisions (6)	Limited quantities (7)	Packagings and IBCs		Portable tanks	
							Packing instruction (8)	Special provisions (9)	Portable tank instruction (10)	Portable tank special provisions (11)
0107	FUZES, DETONATING^	1.2B				NONE	P141			
0110	GRENADES, PRACTICE, hand or rifle^	1.4S				NONE	P141			
0113	GUANYL NITROSAMINOGUANYLIDENE HYDRAZINE, WETTED with not less than 30% water, by mass^	1.1A			266	NONE	P110 (a) or (b)	PP42		
0114	GUANYL NITROSAMINOGUANYLTETRAZENE (TETRAZENE), WETTED with not less than 30% water, or mixture of alcohol and water, by mass^	1.1A			266	NONE	P110 (a) or (b)	PP42		
0118	HEXOLITE (HEXOTOL), dry or wetted with less than 15% water, by mass^	1.1D				NONE	P112			
0121	IGNITERS^	1.1G				NONE	P142			
0124	JET PERFORATING GUNS, CHARGED, oil well, without detonator^	1.1D				NONE	P101			
0129	LEAD AZIDE, WETTED with not less than 20% water, or mixture of alcohol and water, by mass^	1.1A			266	NONE	P110 (a) or (b)	PP42		

0130	LEAD STYPHNATE (LEAD TRINITRORESORCINATE), WETTED with not less than 20% water, or mixture of alcohol and water, by mass^	1.1A		266	NONE	P110 (a) or (b)	PP42
0131	LIGHTERS, FUSE^	1.4S			NONE	P142	
0132	DEFLAGRATING METAL SALTS OF AROMATIC NITRODERIVATIVES, N.O.S.^	1.3C			NONE	P114 (a) or (b)	PP26
0133	MANNITOL HEXANITRATE (NITROMANNITE), WETTED with not less than 40% water, or mixture of alcohol and water, by mass^	1.1D		266	NONE	P112(a)	
0135	MERCURY FULMINATE, WETTED with not less than 20% water, or mixture of alcohol and water, by mass^	1.1A		266	NONE	P110 (a) or (b)	PP42
0136	MINES with bursting charge^	1.1F			NONE	P130	
0137	MINES with bursting charge^	1.1D			NONE	P130 LP101	PP67 L1
0138	MINES with bursting charge^	1.2D			NONE	P130 LP101	PP67 L1
0143	NITROGLYCERIN, DESENSITIZED with not less than 40% non-volatile water-insoluble phlegmatizer, by mass^	1.1D	6.1	266, 271	NONE	P115	PP53 PP54 PP57 PP58
0144	NITROGLYCERIN SOLUTION IN ALCOHOL with more than 1% but not more than 10% nitroglycerin^	1.1D			NONE	P115	PP45 PP55 PP56 PP59 PP60
0146	NITROSTARCH, dry or wetted with less than 20% water, by mass^	1.1D			NONE	P112	

(continued)

Table 11.1 (Cont.)

DANGEROUS GOODS LIST

UN No. (1)	Name and description (2)	Class or division (3)	Subsidiary risk (4)	UN packing group (5)	Special provisions (6)	Limited quantities (7)	Packagings and IBCs		Portable tanks	
							Packing instruction (8)	Special provisions (9)	Portable tank instruction (10)	Portable tank special provisions (11)
0147	NITRO UREA^	1.1D				NONE	P112(b)			
0150	PENTAERYTHRITE TETRANITRATE (PENTAERYTHRITOL TETRANITRATE; PETN), WETTED with not less than 25% water, by mass, or PENTAERYTHRITE TETRANITRATE (PENTAERYTHRITOL TETRANITRATE; PETN), DESENSITIZED with not less than 15% phlegmatizer, by mass^	1.1D			266	NONE	P112(a) or			
0151	PENTOLITE, dry or wetted with less than 15% water, by mass^	1.1D				NONE	P112			
0153	TRINITROANILINE (PICRAMIDE)^	1.1D				NONE	P112 (b) or			
0154	TRINITROPHENOL (PICRIC ACID), dry or wetted with less than 30% water, by mass^	1.1D				NONE	P112 (a), (b)or (c)	PP26		
0155	TRINITROCHLOROBENZENE (PICRYL CHLORIDE)^	1.1D				NONE	P112 (b) or (c)			

0159	POWDER CAKE (POWDER PASTE), WETTED with not less than 25% water, by mass^	1.3C	266	NONE	P111	PP43
0160	POWDER, SMOKELESS^	1.1C		NONE	P114 (b)	PP50 PP52
0161	POWDER, SMOKELESS^	1.3C		NONE	P114 (b)	PP50 PP52
0167	PROJECTILES with bursting charge^	1.1F		NONE	P130	
0168	PROJECTILES with bursting charge^	1.1D		NONE	P130 LP101	PP67 L1
0169	PROJECTILES with bursting charge^	1.2D		NONE	P130 LP101	PP67 L1
0171	AMMUNITION, ILLUMINATING with or without burster, expelling charge or propelling charge^	1.2G		NONE	P130 LP101	PP67 L1
0173	RELEASE DEVICES, EXPLOSIVE^	1.4S		NONE	P134 LP102	
0174	RIVETS, EXPLOSIVE	1.4S		NONE	P134 LP102	
0180	ROCKETS with bursting charge^	1.1F		NONE	P130	
0181	ROCKETS with bursting charge^	1.1E		NONE	P130 LP101	PP67 L1
0182	ROCKETS with bursting charge^	1.2E		NONE	P130 LP101	PP67 L1
0183	ROCKETS with inert head^	1.3C		NONE	P130 LP101	PP67 L1
0186	ROCKET MOTORS^	1.3C		NONE	P130 LP101	PP67 L1
0190	SAMPLES, EXPLOSIVE, other than initiating explosive^		16, 274		P101	

(continued)

Table 11.1 (Cont.)

DANGEROUS GOODS LIST

UN No. (1)	Name and description (2)	Class or division (3)	Subsidiary risk (4)	UN packing group (5)	Special provisions (6)	Limited quantities (7)	Packagings and IBCs		Portable tanks	
							Packing instruction (8)	Special provisions (9)	Portable tank instruction (10)	Portable tank special provisions (11)
0191	SIGNAL DEVICES, HAND^	1.4G				NONE	P135			
0192	SIGNALS, RAILWAY TRACK, EXPLOSIVE^	1.1G				NONE	P135			
0193	SIGNALS, RAILWAY TRACK, EXPLOSIVE^	1.4S				NONE	P135			
0194	SIGNALS, DISTRESS, ship^	1.1G				NONE	P135			
0195	SIGNALS, DISTRESS, ship^	1.3G				NONE	P135			
0196	SIGNALS, SMOKE^	1.1G				NONE	P135			
0197	SIGNALS, SMOKE^	1.4G				NONE	P135			
0204	SOUNDING DEVICES, EXPLOSIVE^	1.2F				NONE	P134 LP102			
0207	TETRANITROANILINE^	1.1D				NONE	P112 (b) or (c)			
0208	TRINITROPHENYLMETHYLNITRA- MINE (TETRYL)^	1.1D				NONE	P112 (b) or (c)			
0209	TRINITROTOLUENE (TNT), dry or wetted with less than 30% water, by mass^	1.1D				NONE	P112 (b) or (c)	PP46		
0212	TRACERS FOR AMMUNITION^	1.3G				NONE	P133	PP69		
0213	TRINITROANISOLE^	1.1D				NONE	P112 (b) or (c)			

0214	TRINITROBENZENE, dry or wetted with less than 30% water, by mass^	1.1D			NONE	P112	
0215	TRINITROBENZOIC ACID, dry or wetted with less than 30% water, by mass^	1.1D			NONE	P112	
0216	TRINITRO-m-CRESOL^	1.1D			NONE	P112 (b) or (c)	PP26
0217	TRINITRONAPHTHALENE^	1.1D			NONE	P112 (b) or (c)	
0218	TRINITROPHENETOLE^	1.1D			NONE	P112 (b) or (c)	
0219	TRINITRORESORCINOL (STYPHNIC ACID), dry or wetted with less than 20% water, or mixture of alcohol and water, by mass^	1.1D			NONE	P112(a), (b) or (c)	PP26
0220	UREA NITRATE, dry or wetted with less than 20% water, by mass^	1.1D			NONE	P112	
0221	WARHEADS, TORPEDO with bursting charge^	1.1D			NONE	P130 LP101	PP67 L1
0222	AMMONIUM NITRATE with more than 0.2% combustible substances, including any organic substance calculated as carbon, to the exclusion of any other added substance^	1.1D			NONE	P112 (b) or (c)	PP47
0224	BARIUM AZIDE, dry or wetted with less than 50% water, by mass^	1.1A	6.1		NONE	P110 (a) or (b)	PP42
0225	BOOSTERS WITH DETONATOR^	1.1B			NONE	P133	PP69
0226	CYCLOTETRAMETHYLENETETRA-NITRAMINE (HMX; OCTOGEN), WETTED with not less than 15% water, by mass^	1.1D		266	NONE	P112(a)	PP45

(continued)

Table 11.1 (Cont.)

DANGEROUS GOODS LIST

UN No. (1)	Name and description (2)	Class or division (3)	Subsidiary risk (4)	UN packing group (5)	Special provisions (6)	Limited quantities (7)	Packagings and IBCs: Packing instruction (8)	Packagings and IBCs: Special provisions (9)	Portable tanks: Portable tank instruction (10)	Portable tanks: Portable tank special provisions (11)
0234	SODIUM DINITRO-o-CRESOLATE, dry or wetted with less than 15% water, by mass^	1.3C				NONE	P114(a) or (b)	PP26		
0235	SODIUM PICRAMATE, dry or wetted with less than 20% water, by mass^	1.3C				NONE	P114 (a) or (b)	PP26		
0236	ZIRCONIUM PICRAMATE, dry or wetted with less than 20% water, by mass^	1.3C				NONE	P114 (a) or (b)	PP26		
0237	CHARGES, SHAPED, FLEXIBLE, LINEAR^	1.4D				NONE	P138			
0238	ROCKETS, LINE-THROWING^	1.2G				NONE	P130			
0240	ROCKETS, LINE-THROWING^	1.3G				NONE	P130			
0241	EXPLOSIVE, BLASTING, TYPE E^	1.1D				NONE	P116 IBC100	PP61 PP62 PP65 B10		
0242	CHARGES, PROPELLING, FOR CANNON^	1.3C				NONE	P130			
0243	AMMUNITION, INCENDIARY, WHITE PHOSPHORUS with burster, expelling charge or propelling charge^	1.2H				NONE	P130 LP101	PP67 L1		

0244	AMMUNITION, INCENDIARY, WHITE PHOSPHORUS with burster, expelling charge or propelling charge^	1.3H	NONE	P130 LP101	PP67 L1
0245	AMMUNITION, SMOKE, WHITE PHOSPHORUS with burster, expelling charge or propelling charge^	1.2H	NONE	P130 LP101	PP67 L1
0246	AMMUNITION, SMOKE, WHITE PHOSPHORUS with burster, expelling charge or propelling charge^	1.3H	NONE	P130 LP101	PP67 L1
0247	AMMUNITION, INCENDIARY, liquid or gel, with burster, expelling charge or propelling charge^	1.3J	NONE	P101	

fundamental to safe carriage of chemicals by sea. Modern measurement instrumentation has achieved an improved flow of information at a consistent standard, and modern control technology permits exact management of operations. However, in order to be able safely to take full advantage of the gains available, there is a need to understand the capability of the instruments and, equally, their limitations. The best source of detailed information about a particular system can be found in the manufacturer's advice, in particular regarding calibration or maintenance requirements. It has been reported that on numerous occasion vessels have loaded chemical cargoes without appropriate Dräger tubes or antidotes for the cargo. It is the vessel Master's responsibility to procure these items as soon as the cargo is fixed. The appropriate tubes and antidote are to be procured before cargo is loaded onboard. A wide range of instrumentation may be fitted on a modern chemical tanker. Only an outline is given here, providing guidance on the safe and efficient operation of the equipment.

FIRST AID WITH REFERENCE TO MSDS

The Material Safety Data Sheet or "MSDS" is an important source of information for personnel working in dangerous conditions. It is one of the three basic elements of the WHMIS (Workplace Hazardous Materials Information System) right-to-know-system. Every MSDS should include the following details: relevant technical information on the substance; a list of its hazardous ingredients (if it is a mixture); chemical hazard data and control measures, such as proper engineering controls and personal protective equipment; instructions in response to accident prevention whilst using the substance, specific handling, storage and disposal procedures; and emergency procedures to follow in the event of an accident. The information provided on the MSDS is required to be comprehensive and must include what can be expected to be known about the material and the hazards it may present. MSDS from different companies may not look the same but they should contain the same basic information. Refer to Figure 11.1 for an indicative sample of the various sections of an MSDS for acetone liquefied refinery gases (LRG) as well as explanation of the corresponding relevant technical information. The order in which sections appear on an MSDS may vary from one supplier to another, but the content of each section is specified by the legislation. Each section of an MSDS must be filled in, even if it only states: "not determined" or "not applicable":

- Section 1 – Material identification,
- Section 2 – Hazardous ingredients,
- Section 3 – Physical data,
- Section 4 – Fire and explosion data,
- Section 5 – Reactivity data,

- Section 6 – Health hazard data,
- Section 7 – First aid measures,
- Section 8 – Preventative measures,
- Section 9 – Storage and handling,
- Section 10 – Spill clean-up and waste disposal and
- Preparation date.

FIRST AID MEASURES

The first aid measures section describes actions to be taken immediately in the event a casualty is accidentally exposed to the hazardous material. The purpose of first aid is to minimise injury and future disability. In serious cases, emergency first aid may be necessary to keep the casualty alive. First aid information needs to be known before working with the material. Often there is no time to find and read the MSDS during an emergency. This means first aid procedures should be periodically reviewed, especially by personnel trained to give emergency first aid. In any case, **all** personnel should know the location of the facilities and equipment for providing first aid; for example, eyewash fountains, safety showers and first aid kits. When onshore medical treatment is necessary, always send the MSDS, if it is readily available, to the emergency facility with the casualty. If the MSDS is not available, send the material's label or a labelled container of the material, if it is small enough to be packaged. Emergency medical responders need to know what the material is and what first aid measures have been carried out. Occasionally, the MSDS may provide additional instructions (or a note to physician) which may be useful to the emergency response team (refer to Figure 11.1).

MEDICAL EMERGENCIES

The vessel must always be prepared for emergency situations related to any cargo operation. Use of safety equipment and proper training for all crew involved is necessary. The MSDS of the cargoes carried to be available onboard. The staff to be trained and familiarised with the countermeasures against accidental contact with personal. The proper PPE to be made available to all personnel. In case of a medical emergency, assistance to be obtained from the vessel's Medical Advisory Service provider. The details are contained in MCCM manual appendix 7. Reference to be made to International Maritime Dangerous Goods (IMDG) Code Supplements, Medical First Aid Guide for use in accidents involving dangerous goods, Emergency Response Procedures for ships carrying dangerous cargoes (EMS Guide) Medical First Aid Guide (MFAG) for use and EMS.

Revision date: 19/12/2023 | Revision: 5 | Supersedes date: 28/04/2016

ReAgent

Visit Our Website
www.reagent.co.uk

CALL: 0800 990 3258 | EMAIL: ENQUIRIES@REAGENT.CO.UK

SAFETY DATA SHEET
ACETONE LRG
According to UK REACH.

SECTION 1: Identification of the substance/mixture and of the company/undertaking

1.1. Product identifier

Product name	ACETONE LRG
Product number	1195
REACH registration number	01-2119471330-49-XXXX
CAS number	67-64-1
EU index number	606-001-00-8
EC number	200-662-2

1.2. Relevant identified uses of the substance or mixture and uses advised against

Identified uses	Solvent. Laboratory reagent.
Uses advised against	Use only for intended applications.

1.3. Details of the supplier of the safety data sheet

Supplier	Reagent Chemical Services 11b - 13 Aston Fields Road Whitehouse Industrial Estate Runcorn Cheshire WA7 3DL T: 01928 716903 (08.30 - 17.00) F: 01928 716425 E: info@reagent.co.uk

1.4. Emergency telephone number

Emergency telephone	OHES Environmental Ltd 24-7 Tel. 0333 333 9939 (24 hour)

SECTION 2: Hazards identification

2.1. Classification of the substance or mixture

Classification (EC 1272/2008)

Physical hazards	Flam. Liq. 2 - H225
Health hazards	Eye Irrit. 2 - H319 STOT SE 3 - H336
Environmental hazards	Not Classified

2.2. Label elements

EC number	200-662-2

Hazard pictograms

1/12

Figure 11.1 Sample Material Safety Data Sheet (MSDS) for acetone.

Chapter 12

Hazard controls

As we discussed in the previous chapter, there are many inherent hazards and dangers lurking in the dark tank bottoms and bilges of oil and chemical tankers. To assist in the avoidance of injury and catastrophic disaster, modern-day tankers can employ a number of methods for managing and reducing the risk of fire and explosion. The four primary methods are inerting, water padding, drying agents and monitoring. Inerting is carried out by filling the cargo tank, associated piping systems and the spaces surrounding the cargo tanks, with a gas or vapour which will not support combustion, and which will not react with the cargo. Equally importantly, this gas or vapour will maintain that condition. Padding is the process whereby the cargo tank and its associated piping systems are filled with a liquid, gas or vapour which separates the cargo from the air and, in so doing, maintains an innocuous condition. Alternatively, drying is a process where the cargo tank and associated piping systems are filled with moisture-free gas or vapour with a dewpoint of −40°C (−40°F) or below at atmospheric pressure. Again, the purpose is to maintain a safe and innocuous condition.

ANTISTATIC MEASURES

Bottom pumping

The loading of bulk liquid chemicals into a ship's tanks often involves the management of static electricity hazards. Controlling the main generators of static electrical fields can eliminate potential threats. The splashing and spraying of static accumulators should be avoided, as they can lead to the formation of charged mists or foams. A charged mist can be ignited even if the temperature does not reach flashpoint. This means splash filling and spraying are a hazard even with high flashpoint cargoes. The filling of a ship's tank should normally be accomplished through a pipeline which ends near the bottom of the tank so that, at least in the earliest stages of loading, the liquid is gently laid on the bottom of the tank. When the rising liquid

DOI: 10.1201/9781003505044-12

covers the pipeline outlet, the turbulence in the tank is reduced meaning fewer static charges are generated.

Safe pumping rate

The faster the liquid flows through the pipeline to the ship's tank, the higher the rate of electrostatic charging. To avoid excessive turbulence within a static accumulator cargo, the velocity of liquid entering a tank should be extremely low until the inlet is well covered. Low velocity also limits any mixing with water that might be present in the tank bottom. After the inlet has been submerged, the flow velocity may be increased, provided turbulence is minimised and the breaking of the liquid surface is avoided.

Presence of water

Most static accumulators are not miscible with water. The presence of water produces two sources of static electricity. First, friction occurs at the surface of the water droplets which are dispersed within the cargo liquid, meaning more static charges are generated than if the cargo liquid did not contain water. Second, the charged droplets settle through the liquid and gather at an interface, producing high voltage at the liquid surface. This process may continue even after tank filling has ceased.

Gas bubbling up through the filled tank

After loading, pipelines are often blown through using air, nitrogen or other gases. When the gas enters the tank from the bottom, it will rise through the liquid in small bubbles, generating a high voltage at the surface. If it is necessary to blow through after loading a static accumulator, the amount of gas permitted to enter the ship's tank should be kept to a practical minimum.

Relaxation time downstream of filters

Micropore filters made of paper, cloth, felt, chamois or metal grid, particularly if deep and thick, are prolific generators of electrostatic charges. Strainers such as perforated metal baskets are not. Where the former are used, the liquid is likely to be highly charged when it leaves the filter in the loading line. For such charges to be relaxed, the liquid should be allowed to flow quietly through the pipeline for a short time, before entering the ship's tank. Practical experience has shown that 30 seconds is usually sufficient. Filters are usually located onshore, so a 30 second transit time is considered adequate, through if the distance between the filter and the ship's tank is not long enough, then either the flow rate should be reduced, the pipe length increased or its diameter enlarged, or a relaxation tank should be provided between the filter and the storage tank.

Unearthed conductors

A conductor having no electrical contact with earth can become charged and rise in voltage through induction (i.e., without physical transfer of charges) and collection (i.e., with physical transfer). An unearthed conductor floating on the surface of a charged liquid actually collects charges from it. A conductor located in a charged mist becomes charged to the same voltage as the mist, even though it cannot collect any charge. In other words, a rise in voltage is possible without charge transfer to an unearthed conductor. If a spark then jumps between the unearthed conductor and an earthed metal surface, all of the energy the former has accumulated will flow instantly into the spark. This provides an increased risk of fire and/or explosion. Avoiding the presence of unearthed conductors in the ship's tanks is therefore of fundamental importance to prevent incendiary sparks, as they provide the electrode from which a spark can jump. Some of the more common types of unearthed conductors found in ship's tanks may include the following:

- The presence of thin metal scraps, including rust. Although they may not float, they can be buoyed up by charged foam,
- Metal couplings at the end of a non-conductive cargo hose used for filling the tank,
- Metal rod(s) or tube(s) attached to gas sampling meters,
- Metal sampling cans or thermometer holders, which are lowered on a non-conductive rope,
- Tank washing machines on the end of a hose having a broken bonding cable, particularly when the hose is empty and/or
- Dropped tools falling through a tank filled with a charged mist from water washing; remember, the mist might not be visible to the naked human eye/detectable by smell.

Projections and probes in tanks

Tanks are sometimes equipped with sounding pipes which extend down from the underdeck towards the liquid surface. Other examples of projections and probes may include high-level alarms, spraying nozzles and fixed tank washing machines. If the liquid being loaded is at a high surface voltage, an incendiary brush discharge to an unbonded projection may take place. The need to avoid such situations will have been considered during the design of the fixed projections inside the cargo tank, and all requirements for safety as to materials of construction, earthing, insulation and static electricity generation will have been checked while the ship was under construction. Even so, it is important that any routine servicing should be performed in accordance with manufacturer's instructions, and no onboard modifications to the equipment itself should be carried out.

Gauging and sampling of tanks

When loading a static accumulator cargo, conductive objects which are not bonded to the ship's structure such as metal sampling cans, gauge tapes and thermometers should not be lowered into the tank. A period of at least 30 minutes should elapse after filling has stopped for the charge to be relaxed before any unbonded metallic or other conductive equipment is introduced to the tank. It should be noted that a metal sounding rod, suspended on a rope, will not be earthed. The metal will become charged when it is immersed in the charged liquid and, when it is then lifted, a metal-to-metal spark may jump between the rod and the rim of the tank opening. This runs a high probability of being incendiary. If the surface voltage of the liquid is exceedingly high, it is possible to generate an incendiary brush discharge to the equipment when it first approaches the surface of the liquid during lowering. Completely non-conductive equipment could in theory be used, but in practice, it is difficult to ensure that such materials remain non-conductive as they are habitually exposed to dirt and moisture. It is therefore best practice to strictly observe the 30 minute (minimum) waiting time. The restriction of this waiting time may be avoided only if the gauging and sampling equipment is lowered inside a sounding pipe that extends all the way down and is connected to the bottom of the tank. This is because the voltage inside the sounding pipe will be small. It is important to note that shorter sounding pipes are inherent unsafe and should be avoided.

Washing of tanks

During tank washing, a charged mist is produced and will remain present throughout the space. This mist will persist for at least a few hours even after the washing has ended. If an unearthed conductor is lowered into the charged mist, it will become charged to a voltage which may be sufficient enough for an incendiary spark to jump to some part of the tank structure. The limitations on water flow rate per nozzle, per machine and per tank should never be exceeded. If the water contains cleaning additives, or is recycled, or the washing medium is other than fresh or clean water, the washing should be conducted in a non-flammable atmosphere, i.e., the tank should be made inert. The practical aspects of tank washing are therefore important to observe.

Steaming

Steam issued from a nozzle will generate a mist of charged water droplets. Therefore, steam should never be injected into a tank that may contain a flammable atmosphere.

Bonding and earthing

An incendiary spark cannot jump between two conductors which are either electrically bonded together or both earthed, because they are kept at the same voltage. Effective bonding is achieved by connecting a metal cable between objects. The cable may be permanently fixed to one conductor and bolted or clamped to the other. At the removable end, contact should be metal to metal and care should be taken to ensure paint, dirt or rust does not hamper the connection. The cable should be sufficiently strong to maintain good resistance to wear and tear. Bonding and earthing cables should be inspected periodically, and their resistance checked with a meter. Many hoses used in marine operations are made electrically conductive. For example, a pair of flanges bolted together will be electrically continuous, as are the flexible joints of metal loading arms. This means bonding or jumping wires around them are not needed. A different electrical phenomenon may be experienced when a tanker is connected to an onshore installation by way of a conductive hose or a metal loading arm. In this instance, the ship, hose, dock and water will form the elements of a battery meaning a large current can flow through the low-resistance hose, even if the voltage difference between the ship and shore is small. This is not static electricity. When the hose is disconnected, the current is suddenly interrupted and an electrical arc may form between the flanges. This poses the risk of igniting any flammable atmosphere existing at the manifold. To prevent such hazards, the ship must be insulated from the shore pipeline by means of an insulating flange or a length of non-conductive hose. This keeps the circuit of the battery open and prevents sparking. However, there is no need to connect the tanker to the dock by a bonding cable, since both are earthed by the water.

VENTILATION

In accordance with the Det Norske Veritas (DNV) rules for ventilation systems within the cargo area, outside the cargo tanks, the ducting used for the ventilation of hazardous spaces should be separate from those used for the ventilation of non-hazardous spaces. Furthermore, ventilation systems within the cargo area should be independent of all other ventilation systems. Air inlets for hazardous enclosed spaces should be taken from areas which, in the absence of the inlet, would be non-hazardous. Air inlets for non-hazardous enclosed spaces should be taken from non-hazardous areas at least 1.5 metres (4.9 feet) from the boundaries of any hazardous area. Where the inlet duct passes through a hazardous space, the duct is to have overpressure relative to the space, unless the mechanical integrity and gas tightness of the duct will ensure gases cannot leak into or from the duct. Air outlets from non-hazardous spaces must be located outside hazardous areas. Air outlets from hazardous enclosed spaces should be located in an open area which, in the absence of the outlet, would be of the same or lesser hazard than the ventilated space.

Ventilation ducts for spaces within the cargo area must not be led through non-hazardous spaces. This means that non-hazardous enclosed spaces should be arranged with ventilation of the overpressure type. Hazardous spaces, however, must have ventilation with under-pressure relative to the adjacent less hazardous space. Starters for fans for ventilation of gas-safe spaces within the cargo area should be located outside this area or on open deck. If electric motors are installed in such spaces, the ventilation capacity must be sufficiently large to prevent the exceedance of temperature limits, taking account of the heat generated by the electric motors. With respect to wire mesh protective screens, the mesh must not measure more than 13 millimetres (0.5 inches) square and must be fitted to the outside opening of the ventilation duct. For ducts where fans are installed, protection screens are also to be fitted inside of the fan to prevent the entrance of foreign objects into the fan housing. Spare parts for fans should always be carried onboard. Normal wear parts for one motor and one impeller are required for each type of fan serving spaces in the cargo area.

Fans serving hazardous spaces

Electric fan motors should not be installed in ventilation ducts for hazardous spaces unless the motor is certified for the same hazard zone as the space served. These fans must be designed with the least possible risk for spark generation. Moreover, minimum safety clearances between the casing and rotating parts should be such so as to prevent any friction from occurring with each other. Under no circumstances is the radial air gap between the impeller and the casing to be less than 0.1 of the diameters of the impeller shaft in way of the bearing, and not less than 2 millimetres (0.07 inches). The air gap need not be more than 13 millimetres (0.5 inches). The parts of the rotating body and of the casing are to be constructed from materials which are recognised as spark proof and having antistatic properties. Furthermore, the installation of the ventilation units onboard should be done in such a way that it would ensure the safe bonding to the hull of the units themselves. Resistance between any point on the surface of the unit and the hull must not be greater than 106 Ohm (Ω). The following combinations of materials and clearances may be used in the manufacturer of the impeller and duct:

- Impellers and/or housing of non-metallic material, with due regard being paid to the elimination of static electricity,
- Impellers and housings of non-ferrous metals, including
    - Impellers of aluminium alloys or magnesium alloys and a ferrous (including austenitic stainless steel) housing on which a ring of suitable thickness of non-ferrous materials is fitted in way of the impeller, due regard being paid to static electricity and corrosion between ring and housing,
    - Impellers and housing of austenitic stainless steel,
    - Any combination of ferrous (including austenitic stainless steel) impellers and housing with not less than 13 millimetre tip design clearance and/or
    - Any combination of an aluminium or magnesium alloy fixed or rotating component, and a ferrous fixed or rotating.

CHEMICAL CARGO SEGREGATION AND COMPATIBILITY

The importance of enforcing the segregation of different chemical cargoes cannot be overstressed. In the case of two or more liquid chemical cargoes which may/will react with one another in a hazardous manner, segregation must be carried out without exception. The product data sheets, together with the International Code for the Construction and Equipment of Ships carrying Dangerous Chemicals in Bulk/ International Bulk Chemical Code (BCH/IBC) Codes, are to be studied carefully to determine the compatibility restrictions when carrying different groups of cargoes. With respect to slops reacting with each other in a hazardous manner, these must not be collected in the same slop tank nor transferred through the same pipes lest they react.

COMPATIBILITY WITH WATER/STOWAGE OF HEATED CARGOES

Some chemical cargoes are not compatible and may even react with water. Therefore, consideration is necessary to avoid stowage of such cargoes adjacent to the water ballast tanks. It is also a requirement that the heating coils be blown through, cleaned and blanked off; alternatively, thermal oil may be used as a heating medium. It is recommended that a cargo to be heated is not stowed adjacent to cargoes which have a low boiling point as the excess evaporation will result in consequent cargo loss and the possible generation of vapour hazards. As a safe margin, the maximum temperature of the heated cargo must be 10°C (50°F) below the boiling point of the unheated cargo. Heated cargoes must never be stowed adjacent to self-reactive cargoes as the excess heating of self-reactive cargoes will shorten the life of the stabilising inhibitor. In these instances, the following artefacts must be provided to the shipper by the manufacturer of the cargo:

- Name and amount of inhibitor added,
- Date the inhibitor was added and the length of its effectiveness and
- The action to be taken should the length of the voyage exceed the effective lifetime of the inhibitor.

The vessel and the vessel charterers must be informed immediately if a product inhibitor certificate is not made available by the shipper or the manufacturer.

COMPATIBILITY WITH THE COATINGS OF THE CARGO TANKS

The suitability of the coating of tanks for loading various chemicals and products must be checked against the coating/paint manufacturer's data sheets before any cargoes are assigned to the vessel's tanks. Also, any temperature limits imposed by the coatings must never be exceeded. Epoxy coatings are capable of absorbing certain chemicals, which may later be released. This could result in the contamination of future cargoes and lead to safety hazards. Similarly, "metal pick-up" from recently applied zinc coatings could contaminate sensitive cargoes.

COMPATIBILITY WITH EDIBLE OILS

Toxic chemicals, as defined by the BCH/IBC Code, must not be carried as the last cargo immediately prior to the carriage of edible oils or else stowed in adjacent tanks sharing common bulkheads with tanks containing edible oils. Likewise, lengths of pipeline serving tanks containing toxic products must never run through tanks containing edible oils, and vice versa. The FOSFA International (Federation of Oils, Seeds and Fats Associations) publication "Operation Procedure for Ocean Carriers of Oil and Fats for Edible and Oleo-Chemical Use" requires that the immediate previous cargo for tanks, lines and pump systems designated to load and transfer fats must have been on the FOSFA International "List of Accepted Previous Cargoes" or not on the FOSFA International "List of Banned Previous Cargoes" currently in force, whichever is appropriate.

CARGO INHIBITION

Procedure for carriage of inhibited flammable chemical products in cargo tanks

In certain conditions of heat, pressure and in the presence of oxygen, some chemical cargo types can become viscous, solid or dense in nature. This self-reaction can cause some cargoes, especially in the presence of high temperatures and oxygen, to initiate an exothermic reaction; in other words,

Figure 12.1 The deck of the chemical tanker BRO ELIZABETH in dry dock in Brest, France.

becoming self-heating resulting in rapidly expansion. This may result in potentially disastrous consequences for the vessel. As a precaution against this, a chemical inhibitor may be added to prevent the cargo from bonding with itself. However, one aspect of inhibitors is that they sometimes require oxygen to activate. This means that the tank cannot be inerted. Always refer to the latest edition of the IBC Code regarding the carriage of inhibited flammable products in cargo tanks of more than 3,000 m^3 and the use of inert gassing. It is also strongly recommended to contact the terminal's cargo manager for advice and guidance before loading potentially hazardous cargoes which may have a chemical inhibitor added. There are many inhibitor types, most of which are toxic and need to be handled with extreme care. Usually, the inhibitor is added by the terminal personnel during the loading programme (refer to Figure 12.1).

Shippers of inhibited cargoes must advise the vessel (and present an inhibitor certificate onboard prior to loading) of the quantity of inhibitor added, the hazards of the inhibitor, the time validity of the inhibitor, the temperature parameters within which the inhibitor will work and the emergency actions to be taken should these parameters be exceeded. The vessel's

Master is to check that the inhibitor validity is sufficient for the anticipated voyage length. The vapour of the cargo will not necessarily contain inhibitor as the two liquids will have differing evaporation properties. Therefore, it is possible for some solid polymer build-up to occur in the tank vents/screens. These must be verified as clear during the voyage and prior to commencing discharge in order to prevent the possibility of damage created by under-pressure in the tanks during the discharge. The temperature of inhibited cargoes must be checked and recorded daily in order to note any abnormal rise that may indicate either inhibitor failure and/or polymerisation. The terminal management office must be advised of any rise in excessive temperature. Inhibited cargoes often need the presence of some oxygen in the tank atmosphere in order to permit the inhibitor to work properly. The minimum level of oxygen is usually stated on the inhibitor certificate but, as a general rule, a cargo containing an inhibitor that needs oxygen should not be carried in an inerted tank. If nitrogen is bubbled through an inhibited cargo (such as when compressed nitrogen is used to clear the cargo hose after loading), the nitrogen will deplete the oxygen dissolved in the liquid, thereby requiring the inhibitor to take oxygen from the atmosphere. This means it is possible that excessive nitrogen used for blowing through might remain and linger in the ullage space.

USCG GUIDE TO THE COMPATIBILITY OF CHEMICALS

The Guide to Compatibility of Chemicals is based in part on information provided to the United States Coast Guard (USCG) by the National Academy of Sciences. The USCG Advisory Committee on Hazardous Materials represents the latest information available to the Coast Guard on chemical compatibility. The accidental mixing of one chemical cargo with another can in some cases be expected to result in a vigorous and hazardous chemical reaction. The generation of toxic gases, the heating, overflow, and rupture of cargo tanks, and fire and explosion are consequences of such reactions. The purpose of the Compatibility Chart is to show chemical combinations believed to be dangerously reactive in the case of accidental mixing. It should be recognised, however, that although the Compatibility Chart provides a broad grouping of chemicals with an extensive variety of possible binary combinations, the Compatibility Chart should therefore not be used as an infallible guide. This is because although one group of chemicals, speaking, can be considered dangerously reactive with another group where an "X" appears on the Chart, there may exist between the groups some combinations which would not dangerously react. It is offered as an aid in the safe loading of bulk chemical cargoes, with the recommendation that proper safeguards be taken to avoid accidental mixing of binary mixtures for which an "X" appears on the Compatibility Chart. Proper safeguards might include consideration of such factors as avoidance of the use of common cargo and vent lines and carriage in adjacent tanks having a common bulkhead. The following procedure explains how the Compatibility Guide should be used in determining compatibility information:

1. Determine the reactivity group of a particular product by referring to the alphabetical list,
2. Enter the Chart with the reactivity group,
3. Proceed across the page. An "X" indicates a reactivity group that forms an unsafe combination with the product in question. For example, crotonaldehyde is listed as belonging to Group 19 (aldehydes) and also has a notation, (2). The Compatibility Chart shows that chemicals in Group 19 should be segregated from sulphuric and nitric acids, caustics, ammonia and all types of amines (aliphatic, alkanol and aromatic).

It is recognised that there are wide variations in the reaction rates of individual chemicals within the broad groupings shown reactive by the Compatibility Chart. Some individual materials in one group will react violently with some of the materials in another group and cause great hazard; others will react slowly, or not at all. Accordingly, a useful addition to the Guide would be the identification of specific materials which might not follow the characteristic reactivities of the rest of the materials in its group. As other exceptions to the Chart become known, they will be listed in subsequent revisions of this manual.

ATMOSPHERIC CONTROL

Ship preparation of the cargo tank atmosphere

For some cargoes, the IBC Code requires vapour spaces within cargo tanks to have specially controlled atmospheres, principally when the cargo is either air reactive resulting in a hazardous situation, or has a low auto-ignition temperature or has a wide flammability range. The correct atmosphere in a tank can be established either by inerting to prevent the formation of flammable mixtures of cargo vapour and air or through padding to prevent chemical reaction between oxygen and the cargo. It may also be necessary to reduce the humidity (dewpoint) of the atmosphere within the cargo system. The extent of atmosphere control to protect the quality of the cargo will normally be specified by the cargo shippers. Some cargoes are extremely sensitive to commercial contamination or discolouration, and for quality control reasons are carried under a blanket of nitrogen that is pure and which must often be obtained from shore.

GAS TESTING

Gas detection equipment

Gas detection equipment is required for ensuring spaces are safe for entry, work or other operations. Their uses include the detection of

- Cargo vapour in air, inert gas or the vapour of another cargo,
- Concentrations of gas in or near the flammable range,
- Concentrations of oxygen in inert gas, cargo vapour or enclosed spaces and
- Toxic gases.

Personnel must fully understand the purpose and limitations of vapour detection equipment, whether fixed or portable. Maintenance records for all gas detection equipment onboard must be maintained by the Chief Officer. Onboard calibration records and shore records are to be maintained together for each meter and are to be updated on each occasion that the instrument is tested or checked. The importance of careful calibration cannot be overemphasised as the gas detection or analysing equipment will only give accurate readings if calibration is carried out strictly in compliance with the manufacturer's instructions and using the correct calibration gases. Where calibration is carried out ashore or by shore technicians, a certificate is to be issued and retained onboard. Instruments must always be checked, zeroed and spanned where applicable before every use as per the manufacturer's instructions. Where calibration is required by the manufacturer's instructions to be carried out ashore or by shore technicians, this must be recorded within the vessel's PMS and all certification issued. In such circumstances at least one unit for each measurement function should remain onboard available for use at all times. Where calibration is carried out ashore or by shore technicians, a certificate is to be issued and retained onboard.

Any equipment not fully operational and/or in good condition, including perished hoses, leaking aspiration bulbs and out-of-date calibration gases or Dräger tubes should be withdrawn from service and reported to the management office. Hoses used with portable gas instruments must be of sufficient length, appropriate to the full depth of the tank or space being tested. Long hoses must be clearly marked at least every 5 metres (16 feet) so that the user can assess the level of the hose in the space. Where the atmosphere testing equipment is not of a uniform manufacture with identical hose fittings, a suitable system is to be created to identify and match the correct hoses with the correct equipment. Hoses compatible with the equipment should be stowed in the same location as the equipment.

OXYGEN ANALYSERS

All ships are supplied with a portable oxygen analyser. This equipment is supplied for use in checking that spaces to be entered have been properly ventilated. It is also to be used on oil and chemical tankers as well as gas carriers, to check that the atmosphere of inerted tanks remains below 7%. Two tests should be carried out on the instrument prior to use and a permanent record of readings kept onboard. These are zero adjustment and span adjustment.

1. Zero adjustment. This is done by using an oxygen-free gas, such as nitrogen or carbon dioxide. Equipment is supplied for this test. Note that CO_2 is paramagnetic and therefore may not give a zero reading on certain instruments.
2. Span adjustment. This must be done in **fresh air** and the instrument carefully checked that the reading has stabilised at 21% before the atmosphere of any space is tested.

The equipment manufacturer's instructions for the particular instrument should be followed carefully to ensure that calibration procedures are correctly carried out. Calibration checks must be carried out every two months.

EXPLOSIMETER

The explosimeter is the name normally associated with the instrument for measuring hydrocarbon gas in air at concentrations below the lower flammable limit (LFL). The technical name is "Catalytic filament combustible gas indicator." A full understanding of the construction and principle of an explosimeter is essential for its safe and efficient use and it is essential that personnel using this instrument carefully study the operating manual. There is also a detailed explanation in the International Safety Guide for Oil Tankers and Terminals (ISGOTT) carried onboard all tankers. The explosimeter measures from 0% to 100% of the lower explosive level (LEL) (1.4% by volume). If the gas-to-air mixture is above the upper explosive limit (6% by volume), the meter reading will initially rise to give a reading of 100% or above but will rapidly fall towards zero because the mixture of gas and air in the combustion chamber is too "rich" to sustain combustion. The meter must therefore be constantly observed for this phenomenon, as a safe reading may be obtained when the atmosphere is in fact highly dangerous. Calibration checks must be carried out at two monthly intervals and when a filament has been changed in accordance with the manufacturer's instructions. Note that, in general, an explosimeter may be calibrated by different gases. It is essential that the correct gas is used otherwise an error may result. Explosimeters will not read hydrocarbon levels in an inert atmosphere.

TANKSCOPE

Although similar to the explosimeter, the "Tankscope" (or Non-Catalytic Heated Filament Gas Indicator) measures hydrocarbons in an inert atmosphere. It indicates their presence as a percentage proportion of the whole atmosphere. The instrument is especially useful during purging with inert gas. It will indicate when the proportion of hydrocarbons has fallen to a level whereby the atmosphere will remain below the LEL on the introduction of fresh air. Calibration checks must be carried out at two monthly intervals.

TOXIC GAS DETECTORS

The toxic gas detector measures low concentrations of toxic gases. Such gases may include carbon monoxide and/or hydrogen sulphide. The type of instrument will normally require a special attachment or tube which the gas is aspirated through. It is necessary to know in advance what gas is expected in order to choose the correct detection tube. The readings are to be compared with the occupational exposure limits or threshold value limits. A minimum list of tube types required for specific vessels is included at the end of this section; however, additional tubes must be carried appropriate to the hazards identified within the Material Safety Data Sheets (MSDS) for the cargo carried.

COMBINED FUNCTION METERS

There are certain instruments which have a combination of functions. Some examples of these equipment include, but are not limited to, the following:

1. Dräger Combiwarn. This instrument measures and monitors flammable vapours as a percentage of the LEL in the range of 0%–50% LEL. It also measures oxygen concentrations. This instrument can be preset to give audible and visual alarms at specific levels.
2. Digiflam 2000. This combines the functions of the Tankscope and an oxygen meter, its main use being the monitoring of crude oil washing (COW) and inert gas operations.
3. Exotox 40. This is supplied specifically for the use in testing and monitoring the atmosphere of enclosed spaces. It combines the functions of an oxygen monitor, explosimeter and a toxic gas monitor

for either carbon monoxide or hydrogen sulphide. It provides continuous monitoring of all three functions and has visual and audible alarms. As with all other instruments, the manufacturer's instructions regarding operation and calibration must be followed at all times.

PERSONAL MONITORING METERS

Some instruments can be carried in a pocket such as the personal oxygen meter (POM), used for entry into enclosed spaces. Such instruments are intended only as a personal monitor and will give an audible and visual alarm if the oxygen content inside the space falls below the preset level. As monitors, they are not designed (and therefore must not be used) for testing the atmosphere for oxygen or other gases. A vessel carrying H_2S cargo must maintain sufficient supply of personal meters to ensure all persons working in the gas-zone are provided with detection equipment. Zero and alarm checks must be made before each use.

SAMPLE LINES

The material and condition of sample lines can affect the accuracy of gas measurements. Sample tubing which is cracked, or blocked, or which has become contaminated with oil or other substances, may seriously affect instrument readings. The tubing must always be checked before and during use and if necessary be cleaned or replaced. It is also important to realise the length of tubing and compare to the meter manufacturer's instructions as to the number of aspirations per metre length. If this is not done, there is a danger that the sample gas may not reach the meter sensor and therefore give a false reading.

Chapter 13

Enclosed spaces and hazardous working

Spaces which are normally entered into during cargo-handling operations, such as the cargo pump rooms and other enclosed spaces which contain cargo-handling equipment, and similar spaces in which work is performed on the cargo, should be fitted with some form of mechanical ventilation system(s) which are capable external operation from outside the enclosed space. Provision should always be made to ventilate such enclosed spaces prior to entering the compartment and operating equipment within the space. In respect of mechanical ventilation inlets and outlets, these should be arranged such to ensure sufficient air movement throughout the enclosed space so as to avoid the accumulation of toxic and/or flammable vapours (taking account of their vapour densities) and to ensure there is sufficient respirable oxygen to provide a safe working environment. As a minimum, the ventilation system should have the capacity to effect no less than 30 changes of respirable air per hour, based on the total volume of the enclosed space. The ventilation system(s) in place should be permanent and normally of the extraction type. Extraction from above and below the floor plates should be possible. In compartments housing motors which drive the cargo pumps, the ventilation should be of the positive-pressure type. Ventilation exhaust ducts from gas-dangerous spaces should discharge upwards in locations at least 10 metres (32.8 feet) in the horizontal direction from the ventilation intake and opening to the ship's accommodation, services and control station spaces and any other gas-safe spaces. Ventilation intakes should be arranged so as to minimise the potential for recycling hazardous vapours from any ventilation discharge opening. Ventilation ducts must not be led through engine rooms, accommodation spaces, working spaces or any other spaces of a similar nature. On a final point, ventilation fans should be approved by the ship's Administration for operation in explosive atmospheres whenever flammable cargoes are carried onboard the vessel.

DOI: 10.1201/9781003505044-13

SPACES NOT NORMALLY ENTERED

Double bottoms, cofferdams, duct keels, pipe tunnels, spaces containing cargo tanks and any other spaces where cargo may accumulate should be capable of ventilation to ensure sufficient respirable air to avoid the accumulation of toxic and/or flammable vapours. They should also ensure sufficient oxygen is available to provide a safe environment prior to entry. Where a permanent ventilation system is not provided for such spaces, approved portable means of mechanical ventilation should be provided in its place. It should also be borne in mind that in all cases the advice given may be subject to local or national regulations and that terminal operators will more often than not have their own safety procedures which could affect cargo handling operations and the measures to be adopted in emergencies. The vessel's Master and all shipboard personnel must be aware of and comply with those regulations and procedures. Their existence will be highlighted by the use of the Ship/Shore Safety Checklist which, together with its guidelines for completion, forms a fundamental part of establishing safe working and operational conditions for the transport of chemicals in bulk by sea.

SAFETY PRECAUTIONS

Inert gas is non-toxic provided a good quality combustion has taken place. Naturally, of course, after its use, there will be a lack of oxygen within the cargo tanks. Oil vapours may also be present. Should it become necessary to enter a tank which has been inerted, the tank should be washed and gas-freed. After this, the entry should not be permitted until a test of the tank atmosphere has been fully carried out and the atmosphere demonstrates there is adequate oxygen and that the concentrate of cargo vapour and carbon monoxide if combustion has been imperfect, has been reduced to a non-toxic level. Prior to anyone entering such a tank, the inert gas system should be checked to ensure that inert gas cannot be accidentally introduced into the tank whilst personnel are inside the tank.

BASIC KNOWLEDGE OF SAFE WORKING PRACTICES AND PROCEDURES IN ACCORDANCE WITH LEGISLATION AND INDUSTRY GUIDELINES RELEVANT TO OIL AND CHEMICAL TANKERS

Precautions to be taken when entering enclosed spaces

The internal structures and workings of modern-day ships are vastly complex with myriad small and enclosed spaces. Many of these enclosed spaces are used for installing machinery or for storing machine parts and/or workshop equipment. Ships also have a matrix of pipelines which run through each of its parts, including enclosed spaces. The main issue of safety arises when personnel have to enter any one of these enclosed places in order to carry out their duties, such as effecting repairs or cleaning. Because of zero ventilation, these enclosed places generate and store toxic gases which are either produced from the chemicals stored in the space or from leaking pipelines. If any member of the ship's crew were to enter such an enclosed place without taking the proper precautions, they would very quickly succumb to toxic poisoning. This more often than not involves unconsciousness followed by death. In order to prevent such avoidable incidents, there are established procedures that must be followed without deviation or exception. Whilst the exact procedures will differ from vessel to vessel, and space to space, a summary of the key safety points is outlined below.

Before entering an enclosed or confined space, a risk assessment must be carried out by a competent officer. The risk assessment must identify whether the space poses an actual or potential risk to health or safety, and if so, what mitigations must be taken to prevent accidents from occurring. This could include the presence of toxic gases or vapours, static electricity hazards, risk of ignition from radio telecommunications equipment etc. In conjunction with the risk assessment, a detailed list of work to be done within the enclosed or confined space should be drawn up. This is used to inform the risk assessment, especially where hazardous duties are to be performed, such as hot work (welding, pipe replacement) or manual washing. The risk assessment must also clearly indicate any emergency and rescue operations to be carried out in the event any personnel inside the enclosed/confined space experience difficulties.

On completion of the risk assessment, a permit to work (PTW) must be applied for and signed off accordingly by a senior member of the ship's deck department. This would ordinarily be either the Master or the Chief Officer. The Chief Engineer may be required to countersign the PTW if required. The PTW must only be valid for a certain time period. Should the time limit expire but further work duties are required, a new PTW must be issued. This requires a new risk assessment and safe working conditions checklist.

Provided the necessary assessments and checklists have been issued, the work gang (i.e., those authorised to enter the space) may begin preparations to effect entry. This includes donning the required PPE, assembling the necessary equipment and/or machinery outside the space and confirming adequate means of communication between the work gang and the spotter are in working order. The spotter is a senior member of the ship's crew who is suitably qualified and experienced person (SQEP) and authorised to oversee the enclosed/confined space work activities. Their responsibilities include ensuring the work gang is properly protected against any hazards whilst inside the space and maintaining constant contact with the work gang. In the event of an accident, the spotter must relay this to the bridge immediately and summon emergency assistance. All personnel entering the space must carry a lifeline which is held by the spotter. The duty officer/ Officer of the Watch (OOW) on the bridge must be regularly informed of progress.

When opening the enclosed space for the first time, precautions must be taken to determine whether the enclosed space is pressurised. All fire hazard potentials should be minimised whenever hot work is to be carried out. This can be accomplished by emptying any fuel tanks or chemical tanks within close proximity to the hot workplace. In the majority of cases, the confined space will need to be thoroughly ventilated before entering. Before making entry, the space has to be checked for its oxygen content and the presence of other gases and/or vapours. This is done by way of an oxygen analyser and gas detector. The oxygen content should read a minimum of 20.8%–21% by volume. Any percentage less than this is not acceptable with more time given for ventilation. Moreover, sufficient means of lighting and/or illumination should be established in the enclosed space before entering. As part of the pre-entry procedures, the "Personnel at Work" signage should be placed in conspicuous locations around the vessel, such as outside the enclosed space, in the engine control room and on the bridge. Where there is the risk that machinery or equipment could cause injury to personnel working inside the enclosed space, these must be isolated and prevented from operation until all enclosed space working is complete and the space is resealed.

Personnel entering enclosed spaces should carry an oxygen analyser with them whenever working inside the enclosed space. This should be fully charged and switched on to monitor the oxygen content inside the space. As soon as the oxygen level drops, the analyser will sound an alarm alerting the wearer that the space must be evacuated with immediate effect. Unless specifically authorised by the Master, and recorded on the PTW, no source of ignition may be taken inside the space at any time. The number of persons entering the space should be constrained to the absolute minimum number of persons needed to carry out the required work safely. This is to limit the number of potential casualties that (1) may need rescuing and (2) may require medical attention. Rescue and resuscitation equipment must be made available outside the confined space, but located such that it does not pose a slip, trip or drop hazard. Rescue equipment includes breathing air apparatus and spare charge bottles, both of which can be heavy and cumbersome. Wherever possible, some means of hoisting an incapacitated person should be readily available.

On completion of the enclosed space work, all personnel are to exit the space without delay. The space should be sealed and checked. Assuming the work task has been completed properly, and the afterwork checklist is filled in, the PTW can be closed out. Once the PTW has been closed out, it must be removed from the active PTW list and stored for future reference in accordance with the SMS standard operating procedure (SOP). The abovementioned procedure is extremely important for safeguarding personnel who are required to enter and work within enclosed and confined spaces. Failure to abide by these, and the ship's own SMS SOP in regard to enclosed space working, can result in severe injury and potential fatalities. Moreover, it is not just the personnel who are working inside the space that is at risk; the same applies to the emergency personnel who may be tasked with retrieving a casualty from within the space.

NEVER PUT YOURSELF, OR OTHERS, AT UNNECESSARY RISK. IF IN DOUBT, STOP!

When making entry into enclosed spaces that are not in normal daily use, great care should be taken to create and maintain safe working conditions, even if the duration of the work is expected to be short. Many fatalities in enclosed spaces have resulted from personnel entering such spaces without proper supervision or adherence to SMS procedures. Sadly, in every case, the fatality would have been avoided if the simple guidance in this chapter had been followed. The rapid rescue of personnel who have collapsed in an enclosed space presents a particular risk. It is a human reaction to go to the aid of a colleague in difficulties, but far too many additional deaths have occurred from impulsive or ill-prepared rescue attempts. The normal oxygen level in fresh air is ~21% by volume. Uncontaminated air with a slightly lower oxygen concentration can be inhaled for some minutes before the effects become apparent. If the oxygen supply to the brain is depleted, the casualty will begin to feel dizzy and experience headaches, before losing consciousness. This is particularly dangerous as the casualty may not recognise that they are in danger or capable of finding their way out of the space. They therefore become a risk to themselves and others. It is well worth noting that in an oxygen-depleted environment, permanent brain damage may occur in as little as 4 minutes. This means a successful rescue depends upon the casualty being removed from the oxygen-depleted environment and thereafter resuscitated in the shortest possible time.

Ensuring a safe atmosphere

When an enclosed space is left sealed and unventilated for any length of time, the internal atmosphere will become unsafe for human life. Either because it contains insufficient oxygen, or because it contains contaminants, or both. The oxygen content can be reduced naturally by the process of rusting or other oxidising conditions, which absorb oxygen from the air, or by the presence of inert gas. Contamination can come from sources such as stores. Decomposition of animal and vegetable oils and fats, a process known as putrefaction (or "going off"), can seriously deplete the oxygen content and evolve toxic gases, making proper ventilation of the space necessary prior to entry. However, it is also possible that oxygen deficiency may be the result of the air in the space being mixed with a contaminant such as cargo vapour. Cargo vapour or inert gas should always be anticipated in cargo tanks, and leakage into adjacent enclosed spaces separated from cargo tanks by a single gas-tight bulkhead should be suspected. Similarly, cargo vapour or inert gas should be anticipated in any space containing cargo handling or inert gas equipment.

It is therefore vital that nobody enters an enclosed space without a breathing apparatus until it has been confirmed that the atmosphere is safe and will remain so. As a general rule, enclosed spaces should not be entered into unless it is considered absolutely necessary. Suitable notices should be prominently displayed to warn and inform personnel about the dangers of entering enclosed spaces. Instructions should clearly explain the precautions to be taken when entering tanks or other enclosed spaces and listing any restrictions placed upon the permitted work. Company procedures should be sufficiently robust that the instructions are followed every time and without deviation or exception. On some ships, there is no door or hatch restricting passage from a pump room into a duct keel. In these circumstances, the duct keel can be regarded as being ventilated by the pump room extractor fans. Nevertheless, the entry of personnel into the duct keel should be subject to a strict safety protocol involving prior notification to a responsible person.

PREPARATIONS PRIOR TO ALLOWING PERSONNEL INTO ENCLOSED SPACES

Prior to allowing personnel to enter an enclosed space, a PTW must be requested, authorised and issued. It is recommended that the PTW is signed only by the Master or the Chief Officer. In some cases, an officer delegated by either may be authorised to sign the PTW on their behalf. This practice should be avoided wherever possible. The PTW should contain a clear indication as to its maximum period of validity (which should not exceed a normal working day), and the maximum time the space can be left unattended (which should not exceed 4 hours). It is critical to ensure that whilst personnel are within an enclosed space the levels of oxygen and any contaminants are frequently monitored and confirmed as within safe limits. If there is any doubt, suitable breathing apparatus and personal protective equipment should be worn, including a lifeline if practicable. The officer responsible for the enclosed space working task should confirm that:

- The space has been thoroughly ventilated by natural or mechanical means to remove any toxic or flammable gases and to ensure an adequate level of oxygen throughout the space,

- All personnel entering the space are professionally trained in enclosed space entry procedures and are familiar with safety and emergency procedures; they should be aware of the ship's procedure for issuing an enclosed space entry permit,
- A trained crew member is standing by at the entrance,
- A reliable system of communication has been established and is understood both by those entering the space and by the crew member standing by at the entrance,
- The appropriate officer of the watch on the bridge in the cargo control room or in the engine room is aware of the enclosed-space entry operations,
- Rescue procedures are in place, and
- Rescue equipment (including lifelines and harnesses) and breathing apparatus are readily available and resuscitation equipment is prepared.

Emergency rescue procedures should clearly set out how to raise the alarm and summon assistance. Access to the space concerned, the deployment of reserve equipment and communication between the emergency party and command centre should also be arranged. In the event of an emergency, under no circumstances should an attending crew member enter the space before help has arrived and the situation has been evaluated, to ensure the safety of those carrying out the rescue operation.

Testing the enclosed space environment prior to entry

Before the space is entered into it should be thoroughly ventilated. The time necessary to ensure thorough ventilation depends on the size of the space, the capacity of the system used, the level of contamination and the efficiency of the ventilation system. Once the space has been ventilated, the atmosphere should be checked accordingly. The oxygen content should be sampled with a suitable and reliable detector: 21% oxygen is required for entry. The principle of measuring the oxygen level in an enclosed space, and interpretation of the figure obtained, must be thoroughly understood. The content of life-sustaining oxygen in standard air is constant at 21%, with the remaining 79% consisting of other gases which are inhalable but do not themselves sustain life. Therefore, confirming that the oxygen level in a compartment is 21% ensures that there is no major component of the atmosphere that is not air. Nevertheless, this may not exclude trace volumes of toxic vapours. If a flammable cargo vapour is/may be present, a combustible gas indicator should also be used. Vapour content as low as practicable, but never more than 1% lower flammable limit (LFL), is required for entry. If a toxic gas is/may be present, the correct toxic gas detector should be used to confirm the level is below the safe operational exposure limit, depending on the nature of the previous contents of the space.

Ventilation should be stopped approximately 10 minutes before tests are made, and not restarted until the tests are completed. Sampling the atmosphere may require the use of breathing apparatus. A number of samples from different locations may have to be taken before the air in the whole space can be judged safe. Readings should be taken at several levels, i.e., at the top, middle and bottom of the space. Suspected vapours which have a relative vapour density greater than that of air will be found at the bottom of the space, and those that have a relative vapour density less than that of air will be found at the top of the space. Vapour will also tend to remain where the ventilating airflow is least effective. Sampling and measurement should be done by personnel trained in the use of the equipment, and sufficiently knowledgeable to understand the results obtained. It is vital that the correct instruments are used. A combustible gas indicator will not measure an oxygen deficiency, nor indicate the presence of toxic gas or the presence of flammable vapour in inert gas. All atmosphere testing equipment used should be of an approved type. It must be correctly maintained, prepared for use in accordance with the manufacturer's guidance and regularly check-tested against standard samples.

Even after a space has been made gas free and found to contain a respirable atmosphere, local pockets of gas should always be anticipated. Cargo residues may be trapped in tank coatings or in residual scale. The generation of vapour should always be considered possible, even after the loose scale has been removed. Hence, all personnel moving around the space, or to different areas of a tank or compartment, or descending to the lower part after work in the upper part, must remain alert to the possible need for further tests to be made. Unless all necessary safety precautions can be followed, spaces should only be entered by personnel wearing breathing apparatus, appropriate protection against exposure to flammable, toxic or corrosive cargo vapours and, if practicable, a lifeline. In chemical tankers, operational entry into cargo tanks may be required before the atmosphere is certified as safe. A documented system should exist to ensure safety throughout any operation when the entry of a contaminated cargo tank, or one suspected of being contaminated, is necessary.

PRECAUTIONS TO BE TAKEN BEFORE AND DURING MAINTENANCE IN GAS DANGEROUS AREAS

Hot work

No hot work may be undertaken inside a compartment until it has been thoroughly ventilated and cleaned. Tests of the atmosphere in the compartment should indicate 21% oxygen content by volume, with flammable vapours as low as possible, but certainly not more than 1% LFL. The compartment must also be free from toxic gases. It is important to continue the ventilation during hot work. No hot work should be undertaken on the open deck unless the area is free from flammable vapour and all compartments (including deck tanks) within a specified radius around the working area have been washed and freed of flammable vapour and/or inerted. Company or national regulations may provide guidance on this distance. If no guidance is available, then the advice in International Safety Guide for Oil Tankers and Terminals (ISGOTT) should be consulted. All sludge, cargo-impregnated scale, sediment or other material likely to give off flammable or toxic vapour, especially when heated, should be removed from an area of at least 10 metres (32 feet) around the hot work. Any combustible material such as insulation should either be removed or protected from the heat source.

Adjacent compartments should either be cleaned, and gas-freed, to hot work standard, or else freed of cargo vapour to not more than 1% LFL and kept inerted, or completely filled with water. No hot work should be undertaken in a compartment beneath any deck tank that is in use. Care should be taken to ensure that no release of flammable vapour or liquid can occur from non-adjacent compartments that are not gas free. An adjacent fuel oil bunker tank may be considered safe if tests using a combustible gas indicator give a reading of not more than 1% LFL in the ullage space of the bunker tank, and no heat transfer through the bulkhead of the bunker tank will be caused by the hot work. No hot work should be carried out on bulkheads of bunker tanks that are in use. All pipelines interconnecting with cargo spaces should be flushed, drained, vented and isolated from the compartment or deck area where hot work will take place. Hot work on pipelines and valves should only be permitted when the item needing repair has been isolated from the system by cold work and the remaining system blanked off. The item to be worked on should be cleaned and gas-freed to a standard that is safe for hot work, regardless of whether or not it is removed from the hazardous cargo area. All other operations utilising the cargo or ballast system should be stopped before hot work is undertaken and throughout the duration of the hot work. If hot work is interrupted for any reason for an extended period, hot work should not be resumed until all precautions have been rechecked and a new hot work permit has been issued.

SAFETY MEASURES FOR HOT AND COLD WORK

Permits to work

Certain safeguards that normally protect the worker may have to be removed when repair or maintenance work is performed. When this occurs, the hazards involved need to be identified and a safe system of work (SSOW) developed to eliminate or control these hazards A PTW or safe work permit (SWP) is a document which identifies the work to be done, the hazard(s) involved and the precautions to be taken. It ensures that all hazards and precautions have been considered before the work begins. PTW/single permit to work (SPW) should always be used when work is performed by an outside agency or employer. The PTW/SPW is a written record that authorises specific work activities, at a specific work location, for a specific time period. Permits are used for controlling and coordinating work tasks to establish and maintain safe working conditions. They ensure that all foreseeable hazards have been considered and that the appropriate precautions are defined and carried out in the correct sequence. The PTW/SPW is an agreement between the issuer and the receiver that clearly and unambiguously documents the conditions, preparations, precautions and limitations that need to be understood before any work commences.

ELECTRICAL SAFETY PRECAUTIONS

When undertaking electrical work, it is important that the correct procedures are followed at all times. This includes the provision and donning of personal protective clothing and the use of protective equipment needed to protect personnel from potential arc flash and shock hazards. These hazards will be identified in the risk assessment. To limit the potential for injury or poor workmanship, any personnel engaged in electrical work must be SQEP and capable of understanding the purpose and function of electrical heat tracing, its electrical power supply/control equipment and how to recognise and avoid the hazards associated with its operation and maintenance. It is equally as important to treat all electrical conductors and circuit parts as though they are energised until such time as they are placed in an electrically safe work condition. This may be achieved by identifying the circuit or equipment to be deenergised and all possible sources of electrical energy supplies to the specific circuit or equipment; and/or interrupting the load currents appropriately, and then opening the circuit disconnecting device(s); and/or visually verifying, wherever possible, that the appropriated circuit disconnecting device is indeed open; and/or applying a lockout/tagout device according to the ship's documented SMS procedure; and/or testing for the absence of voltage with an approved voltmeter (where the voltmeter is tested on a known circuit voltage prior to and immediately following application); and/or grounding the phase conductors or circuit parts before touching them where the possibility of induced voltages or stored electrical energy exists; and/or finally applying ground-connecting devices which are rated for the available fault duty where the conductors or circuit parts being deenergised could possible contact other exposed energised conductors or circuit parts. To limit the risk of accident electrocution, always ensure insulation mats are provided at switchboard and junction box locations and personnel are issued with, and use, certified high-voltage gloves.

Chapter 14

Emergencies, fire safety and firefighting

GENERAL

Loss of stability

If loss of stability becomes evident or is suspected at any time during cargo loading or discharging, the vessel must take appropriate steps to redress the situation. This is particularly important in double-hull ships without centre-line bulkheads. While specific procedures will differ from one vessel to another, the general response actions are the same:

1. Immediately stop all cargo and other operations such as ballast and bunkers. Activate the emergency shut down (ESD) of cargo operations,
2. Inform the terminal operator,
3. Advise the duty officer, Chief Officer and the Master,
4. Ensure all mooring ropes are tight,
5. Carefully check the levels in all slack tanks (ballast, cargo and bunker),
6. Determine the cause (e.g., incorrect or deviation from the loading/discharging plan or technical cause such as valves or other causes of cargo/ballast internal transfer),
7. Enter the vessel's data into the loading computer to check the vessel's GM; check for the angle of lolling and implement any preventive action accordingly,
8. Create a draft plan for correcting the situation. No action is to be taken without the prior permission of the onshore vessel manager, whose responsibilities include obtaining advice from the vessel's appointed damage stability provider. The only exception to this is when the Master considers immediate action is required to save the vessel from further risk,
9. When loss of stability has occurred, on no account is any ballast or cargo to be pumped out. Where ballasting is required, only

 DOI: 10.1201/9781003505044-14

split double-bottom tanks are to be filled, starting with the side which the vessel is listed over to, before making the ship upright with double-bottom tanks on the opposite side. On no account are dirty ballast tanks that run the full width of the ship to be ballasted as this could increase the free-surface effect,

10. Before attempting to correct stability, the plan must be carefully checked using the ship's loading computer in order to check the criteria at every stage of the plan. The plan is to be agreed with the terminal operator before commencing the operation; all hoses are to be disconnected,
11. Once stability is restored, further investigations should be made to ensure that adequate stability is maintained for the remainder of the cargo operation.

OVERSTRESSING DUE TO HIGH-DENSITY CARGOES

If overstressing is caused by high-density cargoes, then the following steps may be taken unless specifically advised otherwise by the vessel's Safety Management System (SMS) and/or Standard Operating Procedures (SOP). It should be noted that as every vessel has a unique design, different plans and/or critical checklists should be adopted:

1. Immediately stop all cargo and other operations, such as ballast and bunkers,
2. Inform the terminal operator,
3. Inform the onshore vessel manager,
4. Ensure all mooring ropes are held tight,
5. Carefully check the levels in all cargo oil tanks, ballast tanks, bunker tanks, freshwater tanks and lubricating oil tanks,
6. Determine the cause (e.g., incorrect or deviation from loading/discharging plan or technical cause such as valves or other cause of cargo/ballast internal transfer),
7. Ensure the vessel data entered into the loading computer is accurate in order to determine the actual stress and stability condition; send the load indicator printout to the onshore vessel manager and the vessel's emergency response service provider,
8. Create a draft plan for correcting the situation. No action is to be taken without the prior permission of the onshore vessel manager whose responsibilities include obtaining advice from the vessel's emergency response service provider. This is particularly important if structural damage has taken place due to overstressing. The

only exception is when the Master considers immediate action is required to avoid further risk or damage from occurring,

9. When overstressing due to high-density cargoes has resulted in structural damage, on no account is any ballast or cargo to be pumped out or cargo adjusted/shifted until confirmation from the onshore vessel manager is received having obtained advice from the vessel's emergency response service provider,
10. Before attempting to rectify the overstressing caused by high-density cargo, the emergency response plan must be carefully checked using the ship's loading computer to check the criteria at every stage of the plan. The plan is to be agreed with the terminal operator before commencing the cargo operation,
11. In the event the vessel has exceeded the tank loading limit as permitted by the vessel's design and construction, the excess cargo must be transferred to other cargo tanks. Where there is no space available in the other ship's tanks, or the tanks are loaded to the maximum permissible load limit, the cargo must be discharged back to the terminal or lightering arrangements made as appropriate,
12. Once the stresses are reduced and brought within acceptable limits, further investigations should be made in order to ensure that the vessel remains within the allowable stress limits (i.e., sea-going) for the remainder of the cargo operation.

Overfilling of cargo tank

All cargo tanks are fitted with a visual and audible high-level alarm which indicates when the liquid level in the cargo tanks approaches the normal full condition (IBC 15.19.6). The high-level alarm system is independent of the overflow control system and the gauging system. The tank overflow system comes into operation when the normal tank loading procedures fail to stop the liquid level from exceeding the normal full condition and provide an audible and visual tank overflow alarm. The high-level alarm, as well as the tank overflow alarm, must be tested prior to each cargo operation and kept on throughout the cargo operation. In the event of overfilling the cargo tank, the following procedures should be followed:

1. Activate the ESD procedure as agreed with the shore/terminal,
2. Inform the terminal operator,
3. Determine the cause (e.g., incorrect or deviation from loading/discharging plan or technical cause such as valves or other cause) which has resulted in overfilling,

4. Check for available slack tanks containing the same cargo or any available empty tanks,
5. Draft a plan for transferring the cargo from the overfilled tank to other suitable cargo tanks which have space to accommodate the cargo ensuring the vessel remains within the allowable limits of stresses and stability at all times,
6. Minimise the trim and ensure the vessel is kept upright,
7. Anti-pollution equipment to be put into a state of readiness and ensure adequate staff are available for implementing anti-pollution measures and clean-up of any cargo spilled on deck/superstructure due to overfilling of the cargo tank,
8. Ensure all personnel are provided with suitable personal protective equipment (PPE) required as per the cargo characteristics and Material Safety Data Sheet (MSDS),
9. The onshore vessel manager is to be kept fully advised of the situation,
10. Once the level in the tank is lowered to within the acceptable range, put in place procedures to avoid similar incidents from occurring.

Polymerisation (solidification of cargo)

Care should be taken to ensure that cargo is sufficiently inhibited to prevent polymerisation at all times during the voyage. When the vessel is carrying cargoes liable to polymerisation, a certificate of inhibition should be provided from the manufacturer of the cargo to the vessel. The certificate of inhibition should clearly list the following information:

1. Name and amount of inhibitor added to the cargo,
2. The date on which the inhibitor was added and the duration of effectiveness,
3. Any temperature limitations that may impact the inhibitor's effective lifespan,
4. The actions to be taken should the period of voyage exceed the effective lifespan of the inhibitor.

The cargo vent systems should be regularly checked for adequacy of operation to avoid blockage from polymer build-up. Polymerised cargoes should not be loaded into spaces that are adjacent to heated cargo tanks. Also, these cargoes should not be handled by cargo lines passing through cargo tanks in which heated cargo is being loaded. In the event of run-off polymerisation, the following procedures may be followed:

1. Inform the terminal manager and onshore vessel manager,
2. Provide the cargo characteristics, stowage plan and the details of the tanks where polymerisation is identified. The details of the cargo adjacent to the tank or tanks where polymerisation has occurred should also be provided (particularly in relation to cargoes that are sensitive to heat),
3. Reduce the temperature of the tank using boundary cooling,
4. Keep the vessel ventilated to avoid over-pressurisation developing in the cargo tanks. This can result in structural damage. **Note:** the cargo vents must be checked and confirmed they are clear and not blocked by polymer build-up,
5. Suitable inhibitors appropriate for the cargo may be used following consultation with the cargo manufacturer,
6. In case of imminent danger, the vessel must refer to the cargo MSDS for the quantity of inhibitor to use and the procedure for adding inhibitor to the cargo,
7. The tank atmosphere and temperatures are to be checked at regular intervals and the terminal, onshore vessel manager and shipper are updated accordingly,
8. All personnel must be provided with suitable PPE as required per the cargo MSDS, and
9. The onshore vessel manager is to be kept fully advised of the situation. It is their responsibility to liaise with the vessel's Class, Flag State, Port State authorities, chemical experts and any other relevant stakeholders such as emergency services.

Brittle fractures

Normal shipbuilding steels rapidly lose their ductility and impact strength below 0°C (32°F). For this reason, care should be taken to prevent cold cargo from coming into contact with such steels, as the resultant rapid cooling can make the metal brittle causing stress due to contraction. In this condition, the metal will be liable to crack. The phenomenon occurs suddenly and is called a "brittle fracture." In the event a brittle fracture occurs, there are a number of procedures to be taken, which are outlined below:

1. Immediately stop all cargo and other operations, such as ballast and bunkers,
2. Inform the terminal operator,

3. Inform the onshore vessel manager,
4. Carefully check the levels in all cargo oil tanks, ballast tanks, bunker tanks, freshwater tanks and lubricating oil tanks,
5. Ensure the data entered into the loading computer is accurate in order to determine the actual stress and stability condition; send the load indicator printout to the terminal operator office and the emergency response service provider (e.g., Lloyds Register Ship Emergency Response [LR SERS], Rapid Response Damage Assessment [RRDA]),
6. The onshore vessel manager is to be kept fully advised of the situation. The majority of tankers have contracts with damage stability providers and assistance from them will be sought in most cases. They will require accurate data on the vessel tank status and weight distribution in order to perform these calculations,
7. Create a draft plan for correcting the situation. No action is to be taken without permission from the onshore vessel manager who will obtain advice from the vessel's emergency response service provider, particularly if structural damage has occurred due to overstressing. The only exception to this is when the Master considers immediate action is required to save the vessel,
8. On no account is any ballast or cargo to be pumped out or cargo adjusted/shifted until confirmation from the onshore vessel manager is received, having obtained advice from the vessel's emergency response service provider. The only exception to this is when the Master considers immediate action is required to save the vessel,
9. In case any cargo has spilled on deck, an immediate response as per the MSDS is to be implemented with all personnel provided appropriate PPE, and
10. If any cargo is released to the sea, anti-pollution measures as per Shipboard Marine Pollution Emergency Plan/Vessel Response Plan (SMPEP/VRP) and the vessel's IAMSAR Manual (MCCM) Manual are to be initiated.

Tank over-pressure

It is a frequent practice at chemical loading ports to control the atmosphere in cargo tanks with nitrogen supplied from shore for the purpose of drying out the vessel's tank and its associated piping system. Nitrogen is also used to purge tanks before loading cargo or padding the cargo in the tank. The nitrogen may be supplied at high pressure (up to 10 bar) and at a high flow rate. Agreement on the procedure for handling the nitrogen is paramount and should be part of the preloading checklist between the ship and shore, with emphasis given on a clear understanding of the transfer rate and pressure. Although the operation is an important stage in cargo handling, it is also potentially hazardous due to the high-pressure gas being introduced into the tank. Some tanks may not be designed to withstand internal pressure, resulting in structural failure at less than 0.5 bar overpressure. The associated risks of the operation should therefore be thoroughly understood. Procedures should be in place to ensure maximum safety throughout the operation, and all personnel involved must be conversant with those procedures.

It is possible to over-pressurise and even rupture a cargo tank if the nitrogen supply from shore is at too high a flow rate or too great a pressure. There have been incidents where structural damage has occurred. Compressed gas is sometimes used by terminals to press products out of shore tanks and into the ship; in these situations, there is an inherent risk of over-pressurisation developing inside the ship's cargo tanks. The gas pressure used for these operations varies but ranges between 2.5 and 5 bar. The point of greatest concern is when the supply into the ship's tank changes from liquid to compressed gas. This can cause an abrupt and dramatic increase in the tank filling rate. Hence, over-pressurisation of a closed tank can occur within seconds, especially when the distance from the manifold to the tank is small or the vapour space in the tank is limited. Over-pressurisation of cargo tanks can result in catastrophic structural failures, explosion hazards and cargo release into the sea or atmosphere. Tank over-pressure can also result from any of the following:

1. Self-heating of cargo due to polymerisation, rollover or handling of sensitive/reactive cargoes,
2. High loading rates (i.e., larger than the vessel's venting capacity or the designed maximum loading rates),
3. Thermal expansion of cargo at sea due to increase of seawater and air temperatures during the voyage,
4. Failure of PV valves, and
5. Clogging of vent pipes/PV valves due to solidification of cargo/polymerisation.

In the event over-pressure develops in any of the ship's cargo tanks, the following procedures may be followed:

1. If alongside a berth/terminal, activate the ESD procedure as agreed with the shore/terminal,
2. Inform the terminal operator (if alongside at berth/terminal),

3. Ventilate the tank via two P/V valves until the pressure is restored to normal. When alongside a berth/terminal, the terminal operator must be advised on the intended/inadvertent release of cargo vapours,
4. If at sea, turn the vessel and adjust the course and speed to bring the wind across the deck, directing the vapours away from the accommodation block,
5. Precautions to be in place as per the MSDS of the cargo. All ship staff are to be provided with appropriate PPE,
6. Implement a total ban on smoking and any hot work, cease any operations which might produce a source of ignition,
7. Investigate the reason for the tank over-pressurisation. This may be due to polymerisation, rollover, sensitive/reactive cargoes, failure of P/V valves, high loading rates, use of compressed gas by shore, introduction of nitrogen into cargo tanks etc.,
8. In the event the tank over-pressurisation has resulted in structural failure, pollution, explosion etc., the guidelines provided in the MCCM Manual, SMPEP and VRP (if in US waters) are to be followed,
9. The onshore vessel manager is to be informed and kept fully advised on the situation,
10. The affected tank must be inspected at the first opportunity for any structural damages caused by over-pressure.

Rollover

Rollover is a spontaneous rapid mixing process which occurs in large tanks as a result of density inversion. Stratification develops when the liquid layer adjacent to a liquid surface becomes denser than the layers beneath. This effect is often caused by the boil-off of lighter fractions from the cargo. Liquid hydrocarbons are most prone to rollover. No external intervention such as vibration, stirring or introducing new liquids is required to initiate "roll-over." The response to small temperature differences within the liquid (which will inevitably occur in the shipboard environment) is sufficient to provide the kinetic energy to start rollover and release the gravitational driving forces which will invert the tank contents. The inversion will be accompanied by a violent evolution of large quantities of vapour and a very real risk of tank over-pressure. If such circumstances are foreseen the tank contents should be circulated daily by the cargo pumps to prevent rollover from occurring. Rollover can also occur if compatible cargoes of different densities are stowed in the same tank. Rollover can be prevented by returning the cargo that is less dense than the bulk liquid on the top of the tank and the cargo that is denser to the bottom of the tank. If two-part cargoes are loaded in the same tank, having a large temperature difference, this will result in a large boil-off and tank over-pressure. Furthermore, rollover can happen on any vessel that has been anchored for an extended period. It is very unlikely – though not unheard of – for rollover to occur during passage. In the event of a rollover, the following procedures may be followed:

1. Activate the ESD procedure as agreed with the shore/terminal,
2. Inform the terminal operator,
3. Inform the onshore vessel manager,
4. Ventilate the tank via two P/V valves until the pressure inside the tank is restored to normal. When alongside a berth/terminal, the terminal operator must be advised on the intended/inadvertent release of cargo vapours,
5. If at sea, turn the vessel and adjust the course and speed to bring the wind across the deck, directing the vapours away from the accommodation block,
6. Precautions to be in place as per the MSDS of the cargo. All ship staff are to be provided with appropriate PPE,
7. Implement a total ban on smoking and any hot work, cease any operations which might produce a source of ignition,
8. Investigate the reason for the tank over-pressurisation. This may be due to polymerisation, rollover, sensitive/reactive cargoes, failure of P/V valves, high loading rates, use of compressed gas by shore, introduction of nitrogen into cargo tanks etc.,
9. In the event the tank over-pressurisation has resulted in structural failure, pollution, explosion etc., the guidelines provided in the MCCM Manual, SMPEP and VRP (if in US waters) are to be followed,
10. The affected tank must be inspected at the first opportunity for any structural damages caused by over-pressure.

Thermal stress

Temperature changes cause the steel structure of the tank to expand or contract. If temperature deformation is permitted to occur freely, no load or stress will be induced on the structure. However, where the temperature deformation is not permitted, an internal stress is created. This internal stress is technically referred to as thermal stress. The main cause of thermal stress is a rapid increase or decrease in temperature. Sensitive/reactive cargoes, run-off polymerisation, rollover etc. can all result in exothermic reactions causing rapid rises in temperature. In the event structural failure or deformation of the strength members/scantlings is found, or suspected, the following procedures should be followed:

1. Immediately cease all cargo and other operations, such as ballast and bunkers,
2. If at the terminal, inform the terminal operator,
3. Carefully check the levels of all cargo oil tanks, ballast tanks, bunker tanks, freshwater tanks and lubricating oil tanks,
4. Ensure the data entered in the loading computer is accurate in order to determine the actual stress and stability condition of the vessel. Send the load indicator printout to the onshore vessel manager and emergency response service provider (e.g., LR SERS and RRDA),
5. Maintain constant communication with the onshore vessel manager. The majority of tankers have contracts with damage stability providers, with assistance sought in most cases. They will require accurate data on the vessel's tank status and weight distribution in order to perform critical safety calculations,
6. Create a draft plan for correcting the situation. No action is to be taken without permission from the onshore vessel manager who will obtain advice from the vessel's emergency response service provider. This is particularly important if structural damage has taken place due to overstressing. The only exception to this is when the Master considers immediate action is required to save the vessel,
7. On no account is any ballast or cargo to be pumped out or cargo adjusted/shifted until confirmation from the onshore vessel manager is received. The only exception to this is when the Master considers immediate action is required to save the vessel,
8. In the event any toxic cargo has spilled on deck or released into the atmosphere, the safety measures of the MSDS are to be followed without deviation,
9. If any structural damage is evident or suspected, the vessel must comply with the emergency procedures contained in the MCCM, SMPEP, and/or VRP (if in US waters).

Emergency discharge due to loss of pumpability

The loss of pumping capability from a tank (either cargo or ballast) can be caused by any of the following conditions:

- Damage to any cargo oil pump (COP),
- Failure of the hydraulic valve system,
- Damage to the hydraulic pipe,
- Blocked pump strainers,

- Ingress of air into the piping and pumping system (causing cavitation), and/or
- Loss of pumpability due to high viscosity or solidification.

In the event the reason for a loss of pumpability is damage to the cargo pump, the following guidelines may be followed, notwithstanding any specific instructions contained in the vessel's SMS and/or SOP:

1. Immediately stop the damaged COP,
2. Inform the terminal,
3. Isolate the damaged cargo pump and operate the other COP using the COP suction and discharge side crossover valves,
4. Due regard must be given when the vessel is carrying incompatible cargoes, ensuring no cargo contamination takes place due to inadequate line and valve segregation,
5. The onshore vessel manager is to be advised should there be any concerns relating to cargo contamination,
6. The shipper is to be informed, prior to using the alternative COP, in case there is a risk of potential cargo contamination or cargo incompatibility, and
7. The reasons for damage to the COP are to be investigated and repairs carried out at the first suitable opportunity.

In the event of failure of the hydraulic valve system, the following procedures may be complied with:

1. Use the emergency hydraulic pump for the operation of the valves,
2. In case it is necessary to stop or reduce the discharge rate, inform the terminal operator immediately,
3. Once the valves are open and lined up as per the cargo plan, restore the discharge rate and advise the terminal,
4. Establish the reasons for the failure of the hydraulic system and carry out the appropriate repairs.

Unable to open the tank valve due to hydraulic line failure

1. In the event the vessel is unable to open the individual tank valve due to a hydraulic line failure, use the stripper valve for discharging the cargo,
2. Ensure the vessel is maintained upright and good trim is maintained throughout,
3. Once the tank level comes down to the stripping level, use the cargo eductors and auto stripping system to ensure the COP does not cavitate,

4. In the event both the main and stripper valves cannot be opened due to hydraulic line failure, inform the terminal operator and the onshore vessel manager to arrange for shore assistance, such as portable pumps for discharging the cargo,

Loss of pumpability due to choked strainer

1. Stop the COP,
2. Inform the terminal operator,
3. Isolate the COP (choked strainer) pump and operate the other COP using the COP suction and discharge side crossover valves,
4. If the vessel is carrying incompatible cargoes, ensure no cargo contamination occurs through inadequate line and valve segregation,
5. Immediately advise the onshore vessel manager if there are concerns of cargo contamination,
6. Inform the shipper and onshore vessel manager prior to using the alternative COP, in case there is any possibility for cargo contamination or cargo incompatibility,
7. In the event the COP (choked strainer) has to be used due to cargo segregation and incompatibility of different grades of cargoes, all cargo operations must be stopped. The COP and the strainer are to be stripped completely prior to opening the strainer cover.
8. The choked strainer is to be thoroughly flushed and cleaned. The strainer cover, along with any other relevant openings and connections, is to be closed and restored as per the original design,
9. Prime up the COP, ensuring all air is removed from the system prior to operating the COP.

Loss of pumpability due to air entering the pipelines/COP (cavitation)

Vessels may suffer loss of pumpability due to air entering the pipelines and the COP caused by leaks in the pipelines, dresser couplings or flanges and leaking valves in tanks. This is most likely to occur when the tanks are emptied or by a vortex effect when discharging cargo at high rates. The following measures may be taken in the event of air ingress into the piping and pumping system:

1. The line and the COP are to be reprimed using the cargo from another tank which has a high liquid column (head),
2. Air is to be removed from the COP and the level separators using the priming cocks provided, and if necessary, by using the auto strip system,
3. In the event the pipeline in use is holed, check for alternate cargo lines, if possible, to discharge (depending on the vessel's pipeline arrangement). Due regard must be given to cargo segregation and compatibility,

4. Start the COP at a slow rate and gradually increase the rate of discharge, ensuring sufficient back pressure is maintained by throttling the COP discharge valve. Try to minimise the potential for vortex effect, which results in air being sucked into the cargo line, by reducing the discharge rate accordingly,
5. Where the cargo in the tank to be pumped out is at an exceptionally low level, use the cargo eductors and/or auto stripping system to pump out the remaining cargo from the tank, and
6. Establish the cause for the pump losing suction and drawing air/cavitation; implement appropriate corrective measures to prevent it from reoccurring.

Loss of pumpability due to high viscosity/solidification of cargo (heated cargoes)

Loss of pumpability can occur if the temperature of the cargo falls below the pour point, particularly for heated cargoes or when discharging cargo in cold weather conditions. In these situations, it has to be borne in mind that for cargoes with a melting point <15°C (59°F), the discharge temperature must be at least 5°C (41°F) above the melting point of the product. For example, benzene has a melting point of approximately 4.5°C (40.1°F). This means it must be discharged at a minimum temperature of 9.5°C (49.1°F) if it is not to solidify. For cargoes with a melting point above 15°C (59°F), the discharge temperature should be at least 10°C (50°F) above the melting point of the cargo. Phenol, for instance, has a melting point of approximately 40.5°C (104.9°F). This means it must be discharged at a minimum temperature of 50.5°C (122.9°F) to prevent it from solidifying. Hence, it may be necessary to increase the heating steam to ensure the recommended temperature for the discharge of cargo is maintained. If the cargo goes below the steam coils and is unpumpable, then the tank must be filled up again to cover the steam coils, reheated and then discharged again. Consideration should be given to minimise the temperature reduction of the heated cargo while taking on ballast in the adjacent tanks. This means ballast intake has to be planned accordingly in order to avoid cargo solidification while carrying heated cargo. The surface area that is exposed between the cargo and the ballast must be minimised. If possible, there should be no exposure of the ballast and cargo via the ship's structure. This is to avoid any cooling effect which ballast water may have on the cargo, leading to solidification. The remaining ballast is to be taken onboard only after the cargo has been fully discharged from the cargo tank. When ballasting double-bottom tanks or side tanks with double bottoms, the level of the ballast should be below the deck head (i.e., the cargo tank bottom) until the cargo is fully discharged.

Portable emergency pump

For vessels fitted with deep well pumps, an additional portable emergency pump is provided for pumping out the cargo in an emergency. The procedure for operating the portable emergency pump is as follows.

Preparation and starting

1. Remove the dust caps from the hydraulic couplings of the return hose and the portable pump,
2. Clean the couplings,
3. Slide the retaining ring and the lock ring backwards,
4. Fit the female coupling over the male coupling and release the retaining ring,
5. Verify that the couplings are properly locked,
6. Connect the pressure hoses accordingly,
7. Similarly, connect the return and the pressure hoses to the main hydraulic line,
8. Remove the dust caps from the couplings of the cargo hose and the portable pump,
9. Clean the couplings,
10. Connect the cargo hose to the pump and one end to the cargo line,
11. Move and install the tripod to a required location above the manhole,
12. Operate the winch on the tripod and slacken the cable to connect to the portable pump,
13. Lower the pump into the tank,
14. Change the switch of the forward hydraulic pump to the cargo pump operation setting,
15. Start the hydraulic pump,
16. Slowly open the hydraulic supply valve on the main hydraulic line pressure side,
17. Increase the rpm of the pump using the flow control valve,
18. To increase the rpm, rotate the valve clockwise.

Monitoring during operation

1. Check the hydraulic pressure and flow,
2. Check the hydraulic hoses regularly for signs of leakage,
3. Check the cargo hose regularly for signs of leakage, and
4. Ensure the portable pump operates smoothly and steadily.

Stopping

1. Stop discharging by turning the flow control valve anticlockwise,
2. Stop the main hydraulic pump,
3. Slowly close the hydraulic supply valve,
4. Hoist the portable pump out of the cargo tank,
5. Disconnect the hydraulic pressure hose from the portable pump and hydraulic system,
6. Disconnect the hydraulic return hose from the portable pump and hydraulic system,
7. Disconnect the discharge hose from the portable pump and cargo line,
8. Clean the portable pump thoroughly before stowing, and
9. Install dust caps on all couplings and ports before securing.

Note: when using the portable emergency pump for discharging toxic/corrosive or flammable cargoes, the precautions stated in the MSDS must be followed without deviation.

SHIP'S ALARMS

The "general emergency alarm" consists of seven or more short blasts followed by one long blast of the ship's horn or whistle (some lines do not sound the signal on the horn or whistle) and by the ship's internal alarm (such as fire alarm bells) accompanied with flashing strobe lights in corridors and public areas for hearing impaired) and PA systems with a tone. For the effective utilisation of the limited emergency equipment available onboard, all personnel must be aware of the location of firefighting gear and lifesaving appliances and be trained in their use. They must also be aware of the alarm signals, easily recognise them and muster at the muster point in case of any type of emergency. The general alarm will be sounded in the event of:

- Fire,
- Collision,
- Grounding,
- Cargo hose burst,
- Major leakage or spillage of oil cargo, and
- Any other event which calls for emergency action.

Other alarms could include:

- Engineer alarm for unmanned machinery spaces,
- Carbon dioxide alarm,
- Fire detector alarms,

- Cargo tank level alarms, and
- Refrigerated store alarm.

If the ship's alarms are ringing, it does not necessarily mean that the situation is out of control. Alarms are warnings, which are sounded so that people onboard take the emergency measures like wearing their life jackets or gathering at a common point, depending upon the type of emergency and instructions given to them. On any vessel, especially oil and chemical tankers, emergencies may have catastrophic consequences, unless prompt and proper action is taken. Actions, therefore, must be deliberate, timely and adequate. Where the opportunity to conduct fire drills with shore establishments exists, these should be taken advantage of to train and develop the crew competence and response capability.

All members of the ship's crew must have, as a minimum, a basic understanding of the principles of fire and how to respond to an outbreak of fire onboard their vessel. The technical personnel onboard must know all the ship emergency codes in detail. This includes understanding the effects of wind, current, shallow water, banks and narrow channels on the ship's behaviour. Every member of the crew should receive appropriate training in accordance with their role in the event of an emergency. Furthermore, any person not a member of the ship's crew and/or passengers onboard must be informed about the possible dangers around the opening and closing of watertight doors, fire doors, valves, scuppers, side scuttles, skylights, portholes and other similar openings. In the unlikely event the Master issues the order to abandon the ship, passengers must be issued lifejackets and evacuated first. The ship's Master should be the last person to leave the ship.

Structure and function of emergency response teams

Every vessel must, in the event of an emergency, have established protocols for an emergency response structure. This typically involves four components: a command centre, the emergency response team itself, a back-up squad and a technical team. Each of these components is responsible for – and therefore tasked with – specific duties. These are briefly outlined below.

Command centre

The command centre is always located on the ship's bridge. The Master, who is responsible for the overall safety and navigation of the ship, will manage and coordinate all communications with the different teams as well as with the shore authorities. A communications log (comms log) must be accurately maintained. The comms log is a legal document

Emergency response team

The emergency response team has the frontline job of responding to the emergency. In general, the Chief Officer will lead the team for the emergency on deck while the Second Engineer will take charge of engine room emergencies. The duties of each member of the emergency response team will be laid down in a formal emergency response plan. Although there are minimum requirements under SOLAS to conduct emergency response practices, it is strongly recommended to carry out tests and practices as often as practicable so as to avoid duplication, confusion and chaos in the event an emergency actually occurs onboard.

Back-up emergency party

The back-up emergency party under the command of an officer should stand by to assist the emergency response team as instructed by the command centre and to provide back-up services, for example, equipment, stores, medical services including cardio-pulmonary resuscitation etc.

Balance crew

The rest of the crew if not allotted any of the duties under the different groups as mentioned above should act as backup for the emergency parties. As backup they may be utilised in various other duties such as accumulating passengers and herding them away from danger to the evacuation decks; escorting feeble passengers or crew including any injured crew to the safe places as designated; rendering first aid and trauma counselling; filling extinguishers as required, mustering fire hoses from elsewhere, recharging and supplying W/T batteries. In the event the Master issues the command to prepare for abandoning the ship, the balance crew may be tasked with taking in additional provisions, clothing/water. Preparation of the survival crafts such that it does not lead to any panic. Making rounds of areas adjacent to the fire area.

Technical team

The technical, or engineer's, team are responsible for maintaining the propulsion and manoeuvring capability of the ship and auxiliary services as far as possible in the circumstances. This group should be under the command of the Chief Engineer or the senior engineering officer onboard and should provide emergency assistance as instructed by the command centre. The prime responsibility for dealing with any emergency in the main machinery spaces will rest with this group. It may be called on to provide additional manpower elsewhere. The plan should ensure that all arrangements apply

Figure 14.1 Sailors wearing firefighting gear enter an office to fight a simulated fire during a general quarters drill.

equally in port and at sea. Duties assigned for the operation of remote controls include:

1. Main engine stop,
2. Ventilation stops,
3. Lubricating and fuel oil transfer pump stops,
4. Dump valves,
5. CO_2 discharge,
6. Watertight door operation, and
7. Operation of essential services such as:
   - Emergency generator and switchboard, and
   - Emergency fire and bilge pumps.

Fire hazards associated with cargo handling and transportation of hazardous and noxious liquids in bulk

The fire hazards associated with noxious liquid substances (NLS) may include, but are not necessarily limited to:

- Cargoes giving off oxygen when on fire, thereby providing the fire with sustenance,

- With some chemical fires, the source of ignition may be heat from a chemical reaction within the cargo itself or through the mixing of the cargo with other chemicals.
- Chemicals that are miscible in fire will render normal foam useless; for such chemicals, alcohol-resistant or dual-purpose foam must be used instead.
- Some chemicals are miscible in water which means their presence may not be recognised,
- Some chemicals are heavier than and insoluble in water; these may be smothered by applying water,
- Some chemicals evolve large volumes of toxic vapours when heated,
- Some chemicals have a low auto-ignition temperature, which means there is a heightened risk of reignition of these chemicals.

Firefighting agents used to extinguish low-fidelity fires

Commonly used firefighting agents include water, foam, carbon dioxide, dry chemical powder (DCP) and alcohol-resistant foam. Water is by far the most commonly used fire extinguishing and cooling agent. This is largely because water possesses excellent heat-absorbing qualities and is available in unlimited quantities at terminals and onboard ships. However, it is **not** suited for the direct extinguishing of oil/chemical-based fires. An alternative to water is foam, which is an aggregation of small bubbles. These have a lower specific gravity than oil or water, which means the foam flows across the surface of a burning liquid forming a coherent smothering blanket. It also helps reduce the surface temperature of the liquid by absorbing much of the heat. A special type of fire extinguishing form is alcohol-resistant foam. This extinguishing agent possesses low-expansion properties and is adaptable to various low-expansion foam generators. Alcohol-resistant foam extinguishing agents may be used onboard chemical tankers where water-soluble flammable liquids such as alcohol, ester, ether, aldehyde, ketone and organic acids are carried. Whereas water and foam agents are excellent for fighting low-fidelity fires such as paper and wood, they are not suited for fires involving electrical equipment, machinery, oils and chemicals. In these situations, an alternative means of extinguishing the fire is needed. One such medium is carbon dioxide.

Carbon dioxide is an excellent smothering agent for extinguishing fires when used in conditions where it will not be widely diffused. This means carbon dioxide is highly effective in enclosed areas such as machinery spaces, pump rooms and electrical switch rooms where it can penetrate into places that cannot be reached by other means. Moreover, unlike water and foam, carbon dioxide does not cause permanent and irreparable damage over and beyond that caused by the fire itself. Just as effective but slightly more damaging than carbon dioxide is DCP. The DCP is discharged from an extinguisher or from a fixed installation as a free-flowing cloud. It is most effective in dealing with fires involving oil spills on a jetty or on the decks of chemical tankers and can also be used in confined spaces. It is especially useful for burning liquids escaping from leaking pipelines and joints.

Firefighting agents used to extinguish chemical fires

The firefighting agents used to extinguish chemical fires must be compatible with the chemical cargoes onboard. This means they must not react with the cargo to form hazardous vapours and reactants. The foam used should also be of an alcohol-resistant aqueous film-forming foam (AR-AFFF) type. AR-AFFF may be used on chemical tankers where water-soluble flammable liquids for example alcohol, ester, ether, aldehyde, ketone and organic acids are carried as cargo. Alternatively, chemical foam may be used as a formation by mixing a solution of alkali (usually sodium bicarbonate), an acid (usually aluminium sulphate), water and a stabiliser. The stabiliser is added to make the foam tenacious and long-lived. When these chemicals react, they form a foam or froth of bubbles which are filled with carbon dioxide gas. The carbon dioxide in the bubbles has little or no extinguishing value. Its only purpose is to inflate the bubbles. Between 7 and 16 volumes of foam can be produced from each volume of water. Premixed foam powders may also be stored in cans and introduced into the water during firefighting operations. For this, a device called a foam hopper is used. Alternatively, the two chemicals may be premixed with water to form an aluminium sulphate solution and a sodium bicarbonate solution.

Fixed firefighting foam operations

All foam systems consist of a water supply, foam liquid storage, a proportioning device and a distribution system. The water supply pump(s) provide(s) a certain capacity of seawater to the deck foam system and is/are supplied by the ship's fire pumps. The foam liquid is stored in a tank. The tank must be complete with a vent, contents gauge and access manhole. The foam is delivered via a high-pressure foam liquid pump to the automatic foam liquid proportionator, which accurately proportions the foam liquid at 3% to 6% to the seawater flow, irrespective of flow rate or pressure. For satisfactory operation of the proportionator, foam liquid must be supplied with a minimum pressure of at least 10 metres (32 feet) head higher than the inlet water pressure under all load conditions. An electrically driven foam liquid pump is provided for this purpose. The foam solution is supplied to the deck monitors and hand lines by the deck main, which is fitted with isolating valves. Each monitor is isolated from the main supply pipe by means of butterfly valves. These are normally closed. Four portable foam-making branch pipes are typically provided, with each branch pipe providing a solution rate of 400 litres per minute.

Portable firefighting foam applicators

Medium expansion foam is most commonly used as the applicator foam. It has an expansion ratio of about 15:1 to 150:1. It is made from the same concentrates as high-expansion foam, but its aeration does not require a fan. Portable applicators can be used to deliver considerable quantities of applicator foam onto spill fires, but as their throw is limited, the foam is liable to be dispersed in moderate winds, they should not be used as the primary firefighting media. Rather, foam applicators are at best a supplement to the deck foam monitors. Sheltered areas not reachable by the foam monitors can be covered by a foam applicator. This provides increased flexibility. Different applicators are available, each offering a varying proportioning ratio. Typically, the applicator needs to be supplied with a fire hose and a foam concentrate container, both of which are stored in the foam station.

Fixed dry chemical systems

DCP is a powder composed of exceedingly small particles usually consisting of sodium bicarbonate, potassium bicarbonate, urea-based potassium bicarbonate or monoammonium phosphate with added particulate material supplemented by special treatment. This provides resistance to packing, resistance to moisture absorption (caking) and proper flow capabilities. DCP works by interrupting the chemical chain reaction sequence as well as providing excellent heat absorption qualities. Multipurpose DCP is usually monoammonium phosphate-based and is effective on fires involving ordinary combustibles, such as wood or paper, as well as fires involving flammable liquids. The main characteristics of DCP are it is best applied to fire extinguishing systems intended for the protection of dangerous and associated artefacts which involve serious hazards and the danger of quick-fire spread. DCP is easy to clean after application and is less contaminant. When responding to fires involving high-tension electrical installations, such as transformers, DCP provides excellent insulation and protection against arcing. Unfortunately, given its nature as a powder, DCP can cause permanent damage when used on equipment and machinery.

Sodium bicarbonate-based dry chemical

This agent consists primarily of sodium bicarbonate ($NaHCO_3$) and is suitable for use on all types of flammable liquid and gas fires (Class B) and also for fires involving energised electrical equipment (Class C). Sodium bicarbonate-based dry chemical is not recommended for the extinguishment of fires in ordinary combustibles (Class A), although it may have a transitory effect in extinguishing surface flaming of such materials.

$$_2NaHCO_3 \rightarrow Na_2CO_3 + CO_2 + H_2O$$

$$Na_2CO_3 \rightarrow Na_2O + CO_2$$

Potassium bicarbonate-based dry chemical

This agent consists primarily of potassium bicarbonate ($KHCO_3$) and is suitable for use on all types of flammable liquid and gas fires (Class B) and also for fires involving energised electrical equipment (Class C). Dry chemicals based on the salts of potassium are not recommended for the extinguishment of fires in ordinary combustible (Class A), although they may have a transitory effect in extinguishing surface flaming of such materials.

$$_2KHCO_3 \rightarrow K_2CO_3 + CO_2 + H_2O$$

$$K_2CO_3 \rightarrow K_2O + CO_2$$

Total flooding system

A total flooding system involves a supply of dry chemicals permanently connected to fixed piping, with fixed nozzles arranged to discharge dry chemicals into an enclosed space or enclosure near the hazard. This type of system may only be used where there is a permanent enclosure about the hazard that is adequate to enable the required concentration to be built up. The leakage of dry chemicals from the protected space should be minimised since the effectiveness of the flooding system depends upon obtaining an extinguishing concentration of dry chemicals. In a total flooding system, the rate of application should be such that the design concentration in all parts of the enclosure is obtained within a maximum of 30 seconds.

Local application system

Local application systems are used for the extinguishment of fires in flammable or combustible liquids, gases and shallow solids such as paint deposits, where the hazard is not enclosed or where the enclosure does not conform to the requirements for total flooding. Application of dry chemicals is usually from nozzles mounted on the tank side or overhead. There are two forms of local application system:

1. Area method, which is applicable to superficial fires. The amount of extinguishing agent depends upon the hazardous area; and

2. Volume method, which is applicable to cubical fires. The amount of extinguishing agent depends on the volume of the object in danger. The hazard shall include all areas that are or may become coated by combustible or flammable liquids or shallow solid coatings, such as areas subject to spillage, leakage, dripping, splashing or condensation and all associated materials or equipment such as freshly coated stock, drainboards, hoods and ducts that might extend fire outside or lead fire into the protected area.

Firefighting operations

Preliminary actions

The individual who discovers the emergency must raise the alarm and pass on information about the situation to the officer on duty who, in turn, must alert the emergency organisation. While this is being put into practice, those on the incident scene should attempt immediate measures to control the emergency until the emergency organisation takes effect. It is worth noting that it is not only cargoes and vapours that pose an existential threat to the ship. For example, galley fires are particularly hazardous as the fumes from burning plastics and cooking oils can quickly overcome the galley staff. Moreover, galley fires can spread easily into the ship's accommodation. The person in charge of the galley or the person first locating the fire should try and extinguish the fire themself after alerting the OOW.

Figure 14.2 Dry chemical agent tanks.

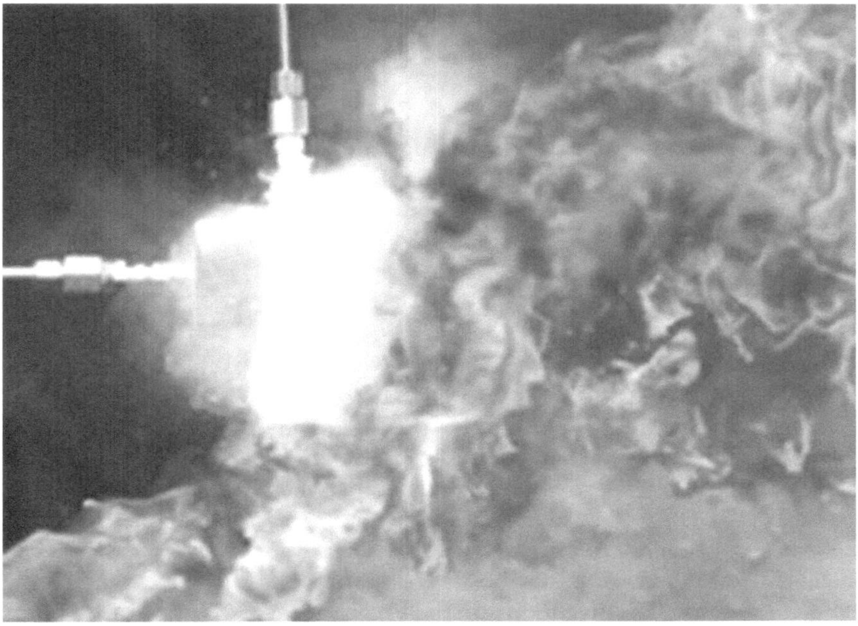

Figure 14.3 Dry chemical agent in operation.

Spill containment in relation to firefighting operations

The first action to be taken in case of an oil/chemical spill and pool fire is the prompt implementation of the ESD. This will do much to limit the amount of liquid spilled and, because of the fish plate, restrict the overflow of cargo overboard. The ESD also works by restricting any potential sources of ignition to ignite the vapour. Once the ESD is fully activated, begin pouring copious amounts of foam gently over the pool fire. The foam will smother and restrict the fire from spreading. Jets of water should **never** be directed onto burning liquid, as this will cause a violent increase in flame and potentially spread the fire. When contained in drip trays, the liquid may also be spilled onto the deck and water; again, water jets should be avoided. In all firefighting operations, full protective clothing must be worn, and all SMS SOPs must be followed without deviation or exception.

Chapter 15

MARPOL and marine oil pollution prevention

The International Convention for the Prevention of Pollution from Ships, or MARPOL Convention, is the main international body of regulations covering the prevention of pollution of the marine environment by ships from operational or accidental causes. It is a combination of two treaties adopted in 1973 and 1978 respectively and updated by amendments throughout the years. The MARPOL Convention was adopted on 2 November 1973 by the IMO and covers pollution by oil, chemicals and harmful substances in packaged form, sewage and rubbish. The Protocol of 1978 relating to the 1973 International Convention for the Prevention of Pollution from Ships (1978 MARPOL Protocol) was later adopted at the Conference on Tanker Safety and Pollution Prevention in February 1978, which was held in response to a spate of tanker accidents between 1976 and 1977. Additional measures relating to tanker design and operation were also incorporated into a Protocol of 1978 relating to the 1974 Convention on the Safety of Life at Sea, 1974, or SOLAS. The Convention includes regulations aimed at preventing and minimising pollution from ships – both from accidental sources and from routine maritime operations – and currently includes six technical Annexes. For the purposes of this book, only Annexes I, II, III and VI are strictly relevant. These four annexes are discussed in brief below. For completeness, Annexes IV and V relate to pollution from sewage and pollution from rubbish, respectively.

SUMMARY OF THE MARPOL ANNEXES I, II, III AND VI

Annex I is the main set of regulations which relate specifically to oil tankers as it includes the regulations for the prevention of pollution by oil. Annex I entered into force on 2 October 1983 and was subsequently revised on 1 January 2007. Annex I covers the prevention of pollution by oil from operational measures as well as from accidental discharges; the 1992 amendments to Annex I made it mandatory for new oil tankers to have double hulls and brought about a phased-in schedule for existing tankers to fit double hulls, which was subsequently revised in 2001 and 2003.

DOI: 10.1201/9781003505044-15

Annex II incorporates the regulations for the control of pollution by noxious liquid substances in bulk. It entered into force on 2 October 1983 and was later revised on 1 January 2007. Annex II details the discharge criteria and measures for the control of pollution by noxious liquid substances carried in bulk; some 250 unique substances are included in a list appended to the Convention; the discharge of their residues is permitted only to reception facilities until certain concentrations and conditions (which vary according to the category and nature of the substance. Under no circumstances can discharges of residues containing noxious substances be permitted within 12 miles (19 kilometres) of the nearest shoreline.

Annex III, or the regulations for the prevention of pollution by harmful substances carried by sea in packaged form, entered into force on 1 July 1992 and contains general requirements for the issuing of detailed standards on packing, marking, labelling, documentation, stowage, quantity limitations, exceptions and notifications for harmful substances carried by sea. For the purpose of Annex III, "harmful substances" are defined as those substances which are recognised as marine pollutants in the International Maritime Dangerous Goods Code (IMDG Code) and/or which meet the criteria provided in the appendix to Annex III of the MARPOL Convention.

Similarly to Annexes IV and V, *Annex VI* or the regulations for the prevention of air pollution from ships applies to all vessels and not just oil and chemical tankers. Annex VI entered into force on 19 May 2005 and set limits on the sulphur oxide and nitrogen oxide emissions from ship exhausts. It also prohibits deliberate emissions of ozone-depleting substances and provides for designated emission control areas (ECA) which are permitted to set more stringent standards for SOx, NOx and particulate matter. In 2011, after extensive work and debate, IMO adopted groundbreaking mandatory technical and operational energy efficiency measures which will significantly reduce the amount of greenhouse gas emissions from ships; these measures were included in Annex VI and entered into force on 1 January 2013.

MARPOL ANNEX I

Oil pollution, as defined by Annex I, is defined as either *waste oil*, which is generated from several areas and systems such as sludge, slop, bilge and the ballast water system; or *intentional discharges* of oil, which can be either legal or illegal discharges. Ship-generated oil waste should be delivered to shore facilities for controlled disposal, or incinerated onboard, or legally or/ and illegally discharged to sea. The effects of oil on marine life are caused by the physical nature of oil (contamination and smothering) or by its chemical components (toxic effect and accumulation leading to tainting). Marine life may also be affected by clean-up operations or indirectly through physical damage to marine habitats. The primary threat to the marine environment is the persistent residue of spilled oils and water-in-oil emulsions ("mousse").

This leads to smothering. The animals and plants most at risk of smothering are those that come into direct contact with a contaminated water surface. These may include marine mammals and reptiles; birds that feed by diving or form flocks on the sea; marine life on shorelines; and animals and plants in aquaculture facilities. The most toxic components in oil tend to be those which are lost rapidly through evaporation when oil is spilt. Because of this, lethal concentrations of toxic components leading to large-scale mortalities of marine life are fortunately relatively rare, tend to be localised and are short-lived. Unfortunately, it is the sublethal effects that impair the ability of marine organisms to reproduce, grow, feed or perform other functions that are necessary for a healthy and diverse marine environment. Sedentary animals in shallow waters such as oysters, mussels and clams that routinely filter large volumes of seawater to extract food are especially prone to accumulate oil components. Whilst these components may not cause any immediate harm, their contamination often renders such animals toxic for human consumption.

Control of operational discharge of oil

Discharges outside special areas. Any discharge of oil or oily mixtures from ships of 400 gross tonnage and above is strictly prohibited except in the limited circumstances when all of the following conditions are met: the ship is proceeding en route; the oily mixture is processed through oil filtering equipment which meets the standards established in Annex I; the oil content of the effluent without dilution does not exceed 15 parts per million; the oily mixture does not originate from cargo pump room bilges on oil tankers; the oily mixture, in the case of oil tankers, is not mixed with oil cargo residues.

Discharges in special areas. Any discharge of oil or oily mixtures from ships of 400 gross tonnage and above are strictly prohibited except when all of the following conditions are met: the ship is proceeding en route; the oily mixture is processed through oil filtering equipment which meets the standards established by Annex I; the oil content of the effluent without dilution does not exceed 15 parts per million; the oily mixture does not originate from the cargo pump room bilges on oil tankers; and the oily mixture, in the case of oil tankers, is not mixed with oil cargo residues.

In respect of the Antarctic area, any discharge into the sea of oil or oily mixtures from any ship is prohibited in **all** circumstances.

Oil filtering equipment

With respect to the provision for processing oily mixtures through oil filtering equipment, any ship of 400 gross tonnage and above must be fitted with oil filtering equipment which complies with the standards established in Annex I. This oil filtering equipment should be of a design approved by the vessel's Flag State Administration and should be such as to ensure that any oily mixture discharged into the sea after passing through the system has an oil content not exceeding 15 parts per million. In addition, it should be provided with an alarm arrangement which indicates when this level cannot be maintained. The system must also be provided with arrangement that ensures any discharge of oily mixtures is automatically stopped when the oil content of the effluent exceeds 15 parts per million.

Oil Record Book, Part I (machinery space operations)

Every oil tanker of 150 gross tonnage and above and every ship of 400 gross tonnage and above other than an oil tanker should be provided with an Oil Record Book Part I (Machinery Space Operations) (ORB, Pt. I). The ORB, whether as a part of the ship's official logbook or otherwise, should be in the form specified in Annex I. The ORB, Pt. I should be completed on each occasion, on a tank-to-tank basis if appropriate, whenever any of the following machinery space operations takes place in the ship:

- Ballasting or cleaning of oil fuel tanks,
- Discharge of dirty ballast or cleaning water from oil fuel tanks,
- Collection and disposal of oil residues (sludge and other oil residues),
- Discharge overboard or disposal otherwise of bilge water which has accumulated in machinery spaces, and
- Bunkering of fuel or bulk lubricating oil.

The ORB, Pt. I must be kept in a location that is readily available for inspection at all reasonable times and, except in the case of unmanned ships under tow, should be kept onboard the ship. Unless local regulations stipulate otherwise, the ORB must be preserved for a period of three years following the final entry.

REQUIREMENTS FOR THE CARGO AREA OF OIL TANKERS

Control of operational discharge of oil

Discharges outside special areas. Any discharge into the sea of oil or oily mixtures from the cargo area of an oil tanker is prohibited except when all the following conditions are satisfied:

- The tanker is not within a special area,
- The tanker is more than 50 nautical miles from the nearest land,
- The tanker is proceeding en route,
- The instantaneous rate of discharge of oil content does not exceed 30 litres per nautical mile,

- The total quantity of oil discharged into the sea does not exceed for tankers delivered on or before 31 December 1979, 1/15,000, of the total quantity of the particular cargo of which the residue formed a part, and for tankers delivered after 31 December 1979, 1/30,000, of the total quantity of the particular cargo of which the residue formed a part, and
- The tanker has in operation an oil discharge monitoring and control system and a slop tank arrangement as required by Annex I.

The provisions of this regulation do not apply to the discharge of clean or segregated ballast.

Discharges in special areas. Any discharge into the sea of oil or oily mixture from the cargo area of an oil tanker is prohibited whilst the vessel is located within a special area. The provisions of this regulation do not apply to the discharge of clean or segregated ballast.

Oil discharge monitoring and control system

Oil tankers of 150 gross tonnage and above must be equipped with an oil discharge monitoring and control system approved by the vessel's Flag State Administration. The system should be fitted with a recording device to provide a continuous record of the discharge in litres per nautical mile and total quantity discharged, or the oil content and rate of discharge. This record should be identifiable as to time and date and must be kept for at least three years. The oil discharge monitoring and control system is required to come into operation whenever there is discharge of effluent into the sea and must be capable of preventing any discharge of oily mixture when the instantaneous rate of discharge of oil exceeds that permitted by Annex I. Any failure of this monitoring and control system must stop the discharge. In the event of failure of the oil discharge monitoring and control system, a manually operated alternative method may be used; however, the defective unit should be made operable as soon as possible.

Oil Record Book, Part II (cargo/ballast operations)

Every oil tanker of 150 gross tonnage and above shall be provided with an Oil Record Book Part II (Cargo/Ballast Operations) (ORB, Pt. II). The ORB, Pt. II, whether as a part of the ship's official logbook or otherwise, must be presented in the format specified in Annex I. The ORB, Pt. II must be completed on each occasion, on a tank-to-tank basis if appropriate, whenever any of the following cargo/ballast operations take place onboard the ship:

- Loading of oil cargo,
- Internal transfer of oil cargo during voyage,
- Unloading of oil cargo,

- Ballasting of cargo tanks and dedicated clean ballast tanks,
- Cleaning of cargo tanks including crude oil washing,
- Discharge of ballast except from segregated ballast tanks,
- Discharge of water from slop tanks,
- Closing of all applicable valves or similar devices after slop tank discharge operations,
- Closing of valves necessary for isolation of dedicated clean ballast tanks from cargo and stripping lines after slop tank discharge operations, and
- Disposal of residues.

Each operation is required to be fully recorded in the ORB, Pt. II so that all entries in the book appropriate to that operation are completed. Each completed operation must be signed by the officer or officers in charge of the operations concerned, with each completed page signed by the Master. The entries in the ORB, Pt. II should be made in either English, French or Spanish. Where entries in an official language of the State whose Flag the ship is entitled to fly are also used, this will usually prevail in the event of dispute or discrepancy. The ORB should be kept in a location that is readily available for inspection at all reasonable times and, except in the case of unmanned ships under tow, must be kept onboard the ship. Furthermore, the ORB must be preserved for a period of three years following the last entry.

Slop tanks

Oil tankers of 150 gross tonnage and above must be provided with slop tank arrangements. In oil tankers delivered on or before 31 December 1979, any cargo tank may be designated as a slop tank. In any case, adequate means must be provided for cleaning the cargo tanks and for transferring the dirty ballast residue and tank washings from the cargo tanks into a slop tank, as approved by the vessel's Flag State Administration. The arrangements of the slop tank or combination of slop tanks must have a capacity necessary to retain the slop generated by tank washings, oil residues and dirty ballast residues. The total capacity of the slop tank or tanks cannot be less than 3% of the oil-carrying capacity of the vessel. For oil tankers of 70,000 tonnes deadweight and above, and delivered after 31 December 1979, at least two slop tanks must be provided onboard.

Pump room bottom protection

This regulation applies to oil tankers of 5,000 gross tonnes deadweight and above constructed on or after 1 January 2007. In accordance with the provisions, the pump room must be provided with a double bottom such that at any cross-section the depth of each double bottom tank or space is such that the distance h between the bottom of the pump room and the ship's base line measured at right angles to the ship's base line is not less than those specified below:

h = B/15(m) or

h = 2 m, whichever is the lesser.

The minimum value of h = 1 m.

Shipboard oil pollution emergency plan (SOPEP)

Every oil tanker of 150 gross tonnage and above and every ship other than oil tankers of 400 gross tonnage and above must carry onboard a shipboard oil pollution emergency plan (SOPEP) that is approved by the vessel's Flag State Administration. The SOPEP is required to include the procedures to be followed by the Master or any other persons having charge of the ship to report an oil pollution incident; the list of authorities or persons to be contacted in the event of an oil pollution incident; a detailed description of the action(s) to be taken immediately by persons onboard to reduce or control the discharge of oil following the incident; and the procedures and point of contact on the ship for coordinating shipboard action with national and local authorities responding to the incident.

Special marine areas

The term "special marine areas" refers to any sea area where for recognised technical reasons in relation to its oceanographical and ecological condition and to the particular character of its traffic, the adoption of special mandatory methods for the prevention of pollution by oil is required. In accordance with the Annex, the primary special marine areas are as follows:

- Mediterranean Sea,
- Baltic Sea,
- Black Sea,
- Red Sea,
- Gulf,
- Gulf of Aden,
- Antarctica and its surrounds,
- Northwest European waters, including the North Sea and its approaches, the Irish Sea and its approaches, the Celtic Sea, the English Channel and its approaches and part of the Northeast Atlantic immediately to the west of Ireland,
- Oman area of the Arabian Sea; and
- Southern South African waters.

RECEPTION FACILITIES

Onshore reception facilities for receiving contaminated and oily water and sludge must be provided in all ports and terminals in which crude oil is loaded into oil tankers where such tankers have immediately prior to arrival completed a ballast voyage of not more than 72 hours or not more than 1,200 nautical miles; all ports and terminals in which oil other than crude oil in bulk is loaded at an average quantity of more than 1,000 tonnes per day; all ports having ship repair yards or tank cleaning facilities; all ports and terminals which handle ships provided with the sludge tank(s); all ports in respect of oily bilge waters and other residues, which cannot be discharged at sea in compliance with Annex I of the MARPOL Convention; and all loading ports for bulk cargoes in respect of oil residues from combination carriers which cannot be discharged at sea in compliance with Annex I of the MARPOL Convention.

REGULATIONS FOR THE CONTROL OF POLLUTION BY NOXIOUS LIQUID SUBSTANCES (NLS)

Substances which pose a threat of harm to the marine environment are divided into three separate categories: X, Y and Z. Category X substances are those which pose the greatest threat to the marine environment, whilst Category Z substances are those posing the least threat. Annex II of the MARPOL Convention prohibits the discharge into the sea of any effluent-containing substances falling under these categories, except when the discharge is made under conditions which are specified in detail for each category. These conditions are prescriptive, and include, where applicable, the maximum quantity of substances per tank which may be discharged into the sea; the speed of the ship during the discharge; the minimum distance from the nearest shoreline during discharge; the minimum water depth at the time of discharge; and the need to effect the discharge below the waterline.

For certain sea areas identified as "special marine areas" more stringent discharge criteria apply. Under Annex II, the sole special marine area is the Antarctic and its surrounds. Annex II requires that every ship be provided with pumping and piping arrangements that ensure that each tank designated for the carriage of Category X, Y and Z substances does not retain after unloading a quantity of residue in excess of the quantity provided in the Annex. For each tank intended for the carriage of such substances, an assessment of the residue quantity must be made. Only when the residue quantity as assessed is less than the quantity prescribed by the Annex may a tank be approved for the carriage of Category X, Y or Z substances. In addition to the conditions referred to above, an important requirement contained in Annex II is that the discharge operations of certain cargo residues and certain tank cleaning and ventilation operations may only be carried out in accordance with approved procedures and arrangements.

To enable the requirement of the above paragraph to be met, a Procedures and Arrangements Manual must be provided and kept onboard the vessel, containing the particulars of the ship's equipment and arrangements in section 2, the operational procedures for cargo unloading and tank stripping in section 3, and in section 4, the procedures for discharge of cargo residues, tank washing, slops collection, ballasting and deballasting as may be applicable to the substances the ship is certified to carry. By following the procedures as set out in this Manual, it will be ensured that the ship complies with all relevant requirements of Annex II to MARPOL 73/78.

CATEGORISATION OF NOXIOUS LIQUID SUBSTANCES AND OTHER SUBSTANCES

For the purpose of the regulations of Annex II, noxious liquid substances are divided into four categories as follows:

1. *Category X:* Noxious Liquid Substances which, if discharged into the sea from tank cleaning or deballasting operations, are deemed to present a major hazard to either marine resources or human health and, therefore, justify the prohibition of the discharge into the marine environment,
2. *Category Y:* Noxious Liquid Substances which, if discharged into the sea from tank cleaning or deballasting operations, are deemed to present a hazard to either marine resources or human health or cause harm to amenities or other legitimate uses of the sea and therefore justify a limitation on the quality and quantity of the discharge into the marine environment,
3. *Category Z:* Noxious Liquid Substances which, if discharged into the sea from tank cleaning or deballasting operations, are deemed to present a minor hazard to either marine resources or human health and therefore justify less stringent restrictions on the quality and quantity of the discharge into the marine environment, and
4. *Other substances:* substances indicated as OS (Other Substances) in the pollution category column of chapter 18 of the International Bulk Chemical Code which have been evaluated and found to fall outside Category X, Y or Z because they are, at present, considered to present no harm to marine resources, human health, amenities or other legitimate uses of the sea when discharged into the sea from tank cleaning of deballasting operations. The discharge of bilge or ballast water or other residues or mixtures containing only

substances referred to as "Other Substances" may not be subject to any requirements of the Annex.

DISCHARGE PROVISIONS

The discharge into the sea of residues of substances assigned to Category X, Y or Z or of those provisionally assessed as such or ballast water, tank washings or other mixtures containing such substances is prohibited unless such discharges are made in full compliance with the applicable operational requirements contained in the Annex. Before any prewash or discharge procedure is carried out in accordance with this regulation, the relevant tank should be emptied to the maximum extent in accordance with the procedures prescribed in the Procedures and Arrangements Manual.

PROCEDURES AND ARRANGEMENTS MANUAL

Every ship certified to carry substances of Category X, Y or Z should carry onboard a Manual approved by the vessel's Flag State Administration. The Manual should have a standard format in compliance with appendix 4 to Annex II. In the case of a ship engaged in international voyages on which the language used is not English, French or Spanish, the text should include a translation into one of these languages. The main purpose of the Manual is to identify for the ship's officers the physical arrangements and all the operational procedures with respect to cargo handling, tank cleaning, slops handling and cargo tank ballasting and deballasting which must be followed in order to comply with the requirements of the Annex.

CARGO RECORD BOOK

Every ship to which Annex II applies must carry some form of cargo record book (CRB), whether as part of the ship's official logbook, or otherwise, in the format specified in appendix 2 to Annex II. After completion of any operation specified in appendix 2 to Annex II, the operation must be promptly recorded in the CRB. Each entry must then be signed by the officer or officers in charge of the operation, with each page signed by the Master. Entries in the CRB are required to be made in either English, French or Spanish. Where entries are made in an official national language of the State whose Flag the ship is entitled to fly are also used, these entries will usually prevail in a dispute or discrepancy. The CRB should be kept in a location that is readily available for inspection and, except in the case of unmanned ships under tow, must be kept onboard the ship. Moreover, the CRB must be retained for a period of three years following the last entry.

SHIPBOARD MARINE POLLUTION EMERGENCY PLAN FOR NOXIOUS LIQUID

Every ship of 150 gross tonnage and above certified to carry noxious liquid substances in bulk must carry onboard a shipboard marine pollution emergency plan for noxious liquid substances approved by the vessel's Flag State Administration. The plan should consist of the procedures to be followed by the Master or other persons having charge of the ship to report a noxious liquid substances pollution incident; a list of authorities or persons to be contacted in the event of a noxious liquid substances pollution incident; a detailed description of the action to be taken immediately by persons onboard to reduce or control the discharge of noxious liquid substances following an incident; and the procedures and point of contact onboard the ship for coordinating shipboard action with national and local authorities.

Consequential amendments to the International Bulk Chemical Code (IBC Code) were also adopted, reflecting the changes to MARPOL Annex II. The amendments incorporate revisions to the categorisation of certain products relating to their properties as potential marine pollutants as well as revisions to ship type and carriage requirements following their evaluation by the Evaluation of Hazardous Substances Working Group. Ships constructed after 1986 carrying substances identified in chapter 17 of the IBC Code must follow the requirements for design, construction, equipment and operation of ships contained in the Code.

ENFORCEMENT OF MARPOL

Any violation of the MARPOL 73/78 Convention within the maritime jurisdiction of any signatory or State which is a party to the MARPOL Convention is punishable under the sovereign laws of the jurisdiction of the vessel's Flag State. In this respect, the term "jurisdiction" is broadly recognised as either the exclusive economic zone for a maritime state or the vessel's Flag State Authority. With the exception of small vessels, ships engaged on international voyages must carry onboard a valid international certificate which may be accepted by foreign port officials as prima facie evidence that the ship complies with the requirements of the MARPOL Convention. If, however, there are clear grounds for believing that the condition of the ship or its equipment does not correspond substantially with the particulars of the certificate, or if the ship does not carry a valid certificate, the Port State Authority carrying out the inspection may detain the ship until the Authority is satisfied that the ship can proceed to sea without presenting an unreasonable threat of harm to the marine environment.

Amendments to the technical Annexes of the MARPOL Convention can be adopted using the "tacit acceptance" procedure, whereby amendments enter into force on a specified date unless an agreed number of State Parties object. In practice, amendments are usually adopted either by the IMO's Marine Environment Protection Committee (MEPC) or by a Conference of Parties to the MARPOL Convention.

Chapter 16

Incident case studies

STOLT VALOR

Synopsis

On 15 March 2012, the chemical tanker STOLT VALOR (15,732 GT; built 2004) (Figure 16.1) carrying 13,000 tonnes of methyl tertiary-butyl ether (MTBE) and 1,300 tonnes of isobutyraldehyde (IBAL) in addition to 430 tonnes of Intermediate Fuel Oil (IFO380) as bunker, suffered an explosion in international waters off Ras Tanura, Kingdom of Saudi Arabia. The vessel's crew were evacuated by the US Navy and shortly after salvors were appointed by the shipowner to respond to the incident. Over the following days, several attempts to tow the vessel away from the coast were made, until the towline broke in bad weather resulting in the STOLT VALOR drifting away from the coast of the Kingdom of Bahrain towards the State of Qatar. At this time, the fire was still raging. A towline was successfully re-established on 19 March a few nautical miles from the coast of the State of Qatar. The casualty was eventually towed offshore.

Response

The fire was controlled on 20 March and an assessment of the vessel was undertaken by salvors. At the time of the inspection, the midship tanks had suffered extensive damage. Boundary cooling of the hotspots and the closing of ventilation and outlets of fuel oil tanks was carried out. Simultaneously, investigations for potential places of refuge, in order to carry out safe removal and lightering of the fuel oil and remaining cargo, were being made with requests sent to the State of Qatar, the Kingdom of Bahrain, the Kingdom of Saudi Arabia and the Islamic Republic of Iran. None of these countries granted refuge to the casualty and bunker oil removal started at sea on 24 March once the risk of further fire and explosion risk were safely reduced. A salvage tug was deployed with oil spill response equipment onboard to

 DOI: 10.1201/9781003505044-16

Figure 16.1 STOLT VALOR.

address any oil spillage during the bunker oil removal operation and daily oil trajectories were simulated as part of the contingency arrangements. Other response capacities were deployed by neighbouring Gulf States at various stages of the incident. The oil removal operation was completed on 2 April with the full cargo removal completed on 29 April. No leakage of oil or cargo was reported throughout the operation and the casualty was eventually towed to a shipyard in the Kingdom of Bahrain to be recycled.

ITOPF involvement

The vessel's owners and P&I Club were in contact with ITOPF[1] from the early stages of the incident and after monitoring remotely it was agreed that ITOPF would travel to the State of Qatar on 19 March in relation to the risk of grounding. ITOPF relocated to the Kingdom of Bahrain on 23 March to assist the salvors and oil spill response contractor. ITOPF liaised extensively with MEMAC (the Marine Emergency Mutual Aid Centre), which played a central coordination role in the incident, as well as with Qatar Petroleum while in the State of Qatar and the Bahraini authorities while in the Kingdom of Bahrain, to discuss the development of the operations, explain the contingency arrangements and to promote a reasonable response from the neighbouring states. A contingency plan as well as risk assessments in relation to the cargo remaining onboard were produced with ITOPF's input at various stages of the response and were presented to the Bahraini authorities and MEMAC. ITOPF also provided comments on a study proposal submitted by the Kingdom of Saudi Arabia and MEMAC to assess the damage caused to the environment as a result of the incident.

Tank explosion case onboard a chemical tanker

Prior to loading the cargo, the vessel was required to gas free and clean her cargo tanks. To accomplish this, the dehumidifier system was used to gas free the tanks. The system consisted of a blower within the forward store, a main line along the deck and branch lines to each cargo tank. Recognising the risk of cargo vapours passing along the line to the forward store, the system was fitted with individual tank valves, a main line isolation valve, a non-return valve and a spool piece, which the shipowner required to be removed when the system was not in operation to isolate the forward store from the cargo side of the system. The day after loading her cargo of Naphtha, and three days prior to the incident, five crew members entered the forward store as part of a familiarisation tour conducted by a junior deck officer. They were met by a strong smell of cargo; however, this was not reported to any of the ship's senior officers.

Two days later and with the vessel now at sea, the ventilation flaps, hatch and door of the forward store were secured. The following day the Bosun opened the door of the forward store and turned on the ventilation fan. An explosion occurred resulting in the death of the Bosun (Figure16.2).

What went wrong (critical factors)

Several valves in the dehumidifier system were passing and the spool piece was not removed after the tank cleaning operation. It was therefore possible for cargo vapours to enter the forward store and with the storeroom secured for sea, an explosive atmosphere was created. The ventilation fan provided the ignition source. Five crew members smelt cargo vapour in the forward store but did not report this to any of the ship's senior officers. Some of the crew members were new to the vessel and junior in rank. The spool piece was not removed as required by the ship's owner. Designed as the last line of defence, it was found that the position and design of the spool piece made it difficult to handle. The spool piece was located 2 metres (6.5 feet) above deck height and weighed in excess 50 kilogram (110 pounds).

Figure 16.2 Exploded forward store.

Lessons learned and recommendations to arise from the STOLT VALOR incident

- Always report any irregularities to a senior officer,
- On completion of operations spool pieces must be removed and the blanks properly fitted. In order to make them conspicuous, all portable bends and spool pieces are to be conspicuously marked and labelled with intended purpose, and
- Vessels with awkward and heavy spool pieces on the dehumidifier system, as was the case onboard the STOLT VALOR, should consider alternative arrangements with the vessel's Superintendent. In response to this incident, the vessel's owners replaced the spool pieces on the dehumidifier system with a flexible hose.

STOLT SKUA INCIDENT

Synopsis

On the morning of 15 April 2012, the chemical tanker STOLT SKUA (Figure 16.3) was on passage from Rotterdam to Antwerp. The ship had

Figure 16.3 STOLT SKUA.

Source: https://commons.wikimedia.org/wiki/File:Stolt_Skua_approaching_Port_of_Rotterdam,_Holland_29-Apr-2007.jpg

finished discharging a cargo of PYGAS in Rotterdam and was completing tank washing and preparation operations prior to loading a new cargo in Antwerp. During the tank cleaning operations, one of the ordinary seaman (OS) was discovered inside one of the tanks being prepared for cargo. As soon as the casualty was discovered, the onboard emergency team were dispatched and quickly arrived on the scene and effected a tank rescue. When recovered from the tank, the OS was unresponsive, and arrangements were made for the OS to be evacuated from the ship. The OS was air lifted by helicopter to the James Paget Hospital in Great Yarmouth. The OS did not regain consciousness and was pronounced dead, at 09:20 hours on 15 April 2012, on arrival at the James Paget Hospital. The cause of death was recorded as "asphyxia in association with inhalation of benzene fumes." The investigation found that the direct cause of the accident was the OS making an unauthorised entry into the cargo tank when a toxic and oxygen-deficient atmosphere was present. Contributing to the accident was a failure to follow the company's procedures for tank cleaning.

Analysis

Supervision of tank cleaning operations by the Chief Officer

In addition to a general safety policy and procedure controlling entry into cargo tanks and other enclosed spaces, Stolt Tankers BV also operate a Tank Cleaning Manual which supplements the other policies and procedures. This Tank Cleaning Manual requires that:

> Whenever a tank cleaning or gas freeing operation is being carried out the either the Chief Officer or Master must be in attendance throughout the operation.

On this occasion, the deck crew were all briefed as to the tank cleaning operation, including the safety precautions to be put in place, and had signed the prepared Tank Cleaning Plan. When the OS joined the tank cleaning team at 06:00 LT, he was also briefed by the Chief Officer as to the operations underway and signed the Tank Cleaning Plan. The briefing by the Chief Officer included a reminder of the absolute prohibition on the entry of any cargo tank until the atmosphere inside the tanks had been tested and an "Entry Permit" issued. Having briefed the tank cleaning crew, the Chief Officer gave the order to issue filter face masks for those entering the tanks. He then retired to his cabin to rest until he would be needed to test the atmosphere of the tanks after venting. This was in direct contravention of the Tank Cleaning Manual in use onboard.

Adequacy of procedures controlling tank cleaning and enclosed space entry onboard STOLT SKUA

As the airborne concentration of benzene during stages of the tank cleaning process could be expected to exceed 1 ppm but was not expected to reach 50 ppm, crew were routinely issued with filter face masks for protection when performing cargo operations on deck when carrying cargoes containing benzene. Amongst other requirements, these procedures required that a tag be permanently attached to the entrance to each tank. These tags were either

- RED: Unsafe for entry,
- GREEN: Tested and safe for entry, all conditions of entry permit met, and
- YELLOW: Tank inerted and unsafe for entry (normally only used when required by shore terminals).

Permits were required for each operation requiring tank entry, and these permits would only be issued after thorough testing of the tank's atmosphere for sufficient oxygen present, the absence of any toxins and the absence of any fire/explosion risk. At the time of the accident a RED (unsafe for entry) tag was prominently displayed at the entrance to 10S tank. Previous records of tank entry were examined and found to comply with the requirements of Stolt Tankers BV's procedures.

Adherence by the deck crew to relevant procedures during the tank cleaning operations

Seawater washing. Seawater was to be introduced into each tank, circulated and then discharged to the sea for a period of 1.5 hours.

Freshwater washing. After the seawater washing was complete, freshwater was to be circulated in the tank for a period of 10 minutes and then discharged to the sea.

Venting. Although the water washing would have diluted and removed any cargo residue remaining in the tank, the atmosphere in the tank would have remained unchanged from when the cargo was discharged from the tank. This atmosphere would consist of vapour from the PYGAS cargo carried and nitrogen introduced into the tank to provide an inert atmosphere. The purpose of venting the tank was to displace this atmosphere with fresh air so that any PYGAS vapour remaining was removed, and sufficient oxygen was introduced into the tank to restore a breathable atmosphere. The tanks were to be initially vented through the ship's vent risers, with the tanks only opened to complete ventilation when the atmosphere in the tanks was expected to be below 30% LEL and the concentration of benzene below the threshold limit value/time weighted average (TLV/TWA) for benzene of 10 ppm. It is then necessary for the tank lid to be opened by the crew to complete the venting. Crew members perform this task wearing filter masks designed to protect against any remaining toxic vapour that may be present in the tank atmosphere.

Atmospheric testing and permit issue. After venting, the atmosphere in each tank was to be tested for the absence of toxic vapour and to confirm the presence of sufficient oxygen. Once the atmosphere in the tank has been confirmed as safe, an "Entry Permit" would be issued allowing crew members to enter the tank.

Ejection. An air-driven Wilden Pump was to be lowered into the tank and used to remove any diluted washings remaining after the sea- and freshwater washing. This pump was to be lowered from deck on ropes and it was not required for any crew to enter the tank to perform this operation.

Mopping and drying. The last stage of the tank cleaning process involved crew members entering the tank to manually mop up any moisture that may remain either in the bottom of the tank or clinging to the inside surfaces of the tank. Once the tank was clean and dry, the next cargo could be loaded without contamination from any previous cargo. Each step in the process was designed to provide the maximum level of safety for those conducting the tank cleaning operations. In particular, although it was not required for a crew member to enter the tank to deploy the Wilden Pump, it would be necessary for them to lean over the open tank hatch and lower the pump to the bottom of the tank. In addition to the crew member wearing a filter mask to protect against toxic vapours, the process was designed to minimise the exposure to any such vapours that may be present. This stage of the process was only to be performed after the atmosphere in the tank had been confirmed as safe.

Rather than waiting for the atmosphere in the tank to be confirmed as safe, the "Optimising Violation" was to lower the Wilden Pump into the tank before the atmosphere had been checked, with the seafarer remaining on deck and not entering the tank. It was not possible to determine if this practice had become the norm onboard STOLT SKUA or if it was employed only in certain circumstances. It was also not possible to determine if the practice was undertaken with the knowledge and approval of the Chief Officer. However, it is clear that this deviation from the standard tank cleaning procedure was being employed on the day of the accident.

Actions of the OS

The post-mortem conducted on the ordinary seaman (OS) showed no evidence of him falling into the tank from the deck onto the first platform on which he was found (a height of 3.31 metres [10.8 feet]). It is therefore concluded that he climbed into the tank and down the ladder onto the first platform. The reason for him to take such a course of action is that he was attempting to move the Wilden Pump to a new location in the tank after it had become caught in the tank's internal structure. Several factors acting in combination may have led the OS to act in contravention of his training and accepted safe working practice; these factors may have included the following:

- Risk taking – taking an action where the outcome is uncertain, often in contravention of norms, regulations or procedures. "I'll take a chance."
- Impulsiveness – inclined to act on impulse rather than thought. "I know what I am doing."

- Invulnerability – impervious to danger or risk. "It won't happen to me."

The tank cleaning operation was already in violation of standard procedures and the OS may have considered his actions in entering the tank as merely a "stretching" of the "accepted violation" already in progress. Routine violations do not necessarily result in accident themselves. Accidents occur when a routine violation occurs in conjunction with an "error." In this case the "error" was the OS's decision to enter the tank.

Conclusions

1. The primary cause of this accident was the OS's decision to enter cargo tank 10S which contained a toxic and oxygen-deficient atmosphere incapable of sustaining life.
2. Contributing to this accident was the practice onboard of not following the established tank cleaning procedures in the belief that doing so would improve tank cleaning performance. A so-called Optimising Violation.
3. The level of supervision and control by the Chief Officer onboard was insufficient to prevent this Optimising Violation occurring.
4. Stolt Tankers BV had robust and adequate procedures in place to prevent such an accident occurring. These procedures met all international requirements and represented "best practice"; however, they were not always followed by the crew of STOLT SKUA.
5. The members of the onboard rescue team acted on knowledge and training, not on emotion and instinct which has led to many failed rescue attempts in the past. The prompt actions of the crew gave the OS every chance of survival.

Actions taken

Following this accident, Stolt Tankers have taken the following actions.

Filter masks

Stolt Tankers have introduced an immediate and comprehensive ban on the use of filter masks on all company vessels.

Replacement equipment

The company has provided each ship with lightweight breathing apparatus sets specifically designed for respiratory protection of the crew while handling toxic cargo on deck. Where possible, the use of Wilden Pumps has been supplemented by the fitting of an eductor on deck.

Awareness raising

Additional safety committee meetings have been held on all ships highlighting the circumstances of this accident and the lessons to be learned.

Training

Stolt Tankers Bureau Veritas (BV) use their investigation into this accident as a case study for training purposes at their "Officers and Crew Safety Excellence" conferences and in other company training courses.

Note

1 www.itopf.org/

Annexes

CARGO TRANSFER PROCEDURE

CHEMICAL TANKER MANUAL	Prepared by	Marine Supdt	
	Checked by	Director	
	Edition	Edition No.4	
	Issue date	Mar 24	
	Approved by	Director	
	Rev date	-	
	Page	1 of 8	

Procedure document 1
Cargo transfer procedure
Dated 20240311

APP 1.1 OIL TRANSFER OPERATIONS AS PER US 33CFR 155.750 (A)(3)/(4)/(5)

App 1.2 GENERAL

These standing orders apply to all personnel engaged in oil transfer operation. In addition to loading and discharging of oil cargoes, transfer operation shall be understood to include ballasting and de-ballasting, tank cleaning and preparation of cargo tanks for loading cargo as well as Crude Oil Washing. The following standing orders have been drawn up so as to comply with United States Coast Guard regulations. A copy of these standing orders shall be permanently maintained in the cargo control room and shall be available for inspection by USCG, Port Officials and ship's personnel.

App 1.3 RESPONSIBILITY OF THE MASTER

The Master shall hold overall responsibility for the Safety and Security of the personnel, vessel and cargo under his command while at the same time ensuring that the vessel complies with all environmental regulations. He is responsible for ensuring that all regulations laid down by the Company, Local Authorities, Governments and International Organisations are fully complied with at all times.

App 1.4 RESPONSIBILITIES OF THE CHIEF OFFICER

The Chief Officer is directly responsible to the Master for the safety of the oil transfer system and the safety of the oil transfer operation. Personnel delegated for the duty watchkeeping during cargo oil transfer operations are directly responsible to the Chief Officer and shall discharge such orders as instructed by the Chief Officer. Tests for integrity of the cargo oil transfer system shall be periodically carried out as laid down in the Code of Federal Regulations (CFR33). Any shortcomings are to be reported to the Master immediately. The Chief Officer along with the Chief Engineer is directly responsible for the correct and safe operation and use of the vessel's inert gas system.

The Chief Officer is to ensure that IGS is in continuous operation during transfer operations as deemed necessary and that the atmosphere in the cargo tanks is maintained under 5% of O_2 and at a positive pressure above 100 mm WG at all times. The Chief Officer is responsible for safe execution of crude oil washing operations whenever this is to be effected during discharge of crude oils. The Chief Officer shall draw up a detailed program prior to each crude oil washing operation and submit same to the Master and the terminal. Guidelines for COW programs are given in the vessel's crude oil washing manual. As the Deputy Ship Security Officer, he is to ensure that the vessel complies with the Security requirements as given in the Company's Ship Security Plan.

App 1.5 RESPONSIBILITY OF THE CHIEF ENGINEER

The Chief Engineer is responsible for ensuring that all engine and deck machinery is in good working order and condition. He is also responsible for the safe and smooth execution of all fuel oil, diesel, gas oil and lubricating oil bunker transfer operations. For details on bunker procedures, see SBM II. He is to ensure that there is close cooperation between the deck and engine departments in order that all operations are carried out in a smooth manner. If he is also the Ship Security Officer, then he is to ensure that the vessel complies with the Security requirements as given in the Company's Ship Security Plan.

	CHEMICAL TANKER MANUAL	Prepared by	Marine Supdt		
		Checked by	Director		
		Edition	Edition No.4		
		Issue date	Mar 24		
		Approved by	Director		
		Rev date	-		
		Page	2	of	8

App 1.6 DELEGATION OF DUTIES AND RESPONSIBILITIES

Salient watch keeping instructions

The Duty Officer holds responsibility for his watch and shall carry out all instructions as directed by the Chief Officer. Any irregularities, shortcomings, defects and similar occurrences shall be brought to the immediate attention of the Chief Officer. No oil transfer shall commence prior to all items of the safety check list being read and fully understood. Sea valves, overboard valves and manifold valves shall never be operated by the duty watch personnel unless expressly instructed by the Chief Officer, except shut-down of same if deemed necessary in an emergency. The Duty Officer shall ensure continuous and effective deck and manifold watches are maintained taking patrol rounds himself whenever the central cargo operations room is manned by Chief Officer.

The Duty Officer shall delegate one of his watch ratings to maintain continuous watch at the cargo manifold. The rating should be provided with means of communication (i.e., VHF/UHF radio), whilst the other rating of his watch shall maintain continuous deck patrol. When the vessel is on a SPM a continuous watch for the Buoy to be kept at the Forecastle Deck. The deck patrol rating shall also be equipped with means of communication (VHF/UHF radio). This rating shall ensure vessel's mooring lines are in order. Periodic reporting system between members of the duty watch shall be drawn up at the beginning of each watch and then effected throughout the watch.

App 1.7 OIL TRANSFER PROCEDURE 33 CFR 155.750

1. This vessel is mainly involved in the carriage of Methanol. Generic and safety information for grade as per the attached data.

2. This vessel is a standard chemical/oil tanker. There are 20 full segregated cargo tanks and 2 slop tanks. Each tank equipped with separate FRAMO pump and each tank pairs (as 1P and 1S, 2P and 2S, etc.) equipped with separate pipelines. Tanks and pumps can be interconnected by various loops on manifold and with port and stbd common lines. Schematic diagrams for pipelines are located in the cargo control room.

3. This vessel has a segregated ballast system independent of cargo pumps and lines. SBT tanks: Forepeak, 1WP, 1WS, 2WP, 2WS, 3C, 3WP, 3WS, 4C, 4WP, 4WS, 5C, 5WP, 5WS, 6WP, 6WS and Aft Peak. Ballasting/ deballasting can be done concurrent with cargo operations.

4. Manifold containment is in drip trays which can be drained into slop tanks. Cargo overflows / leaks onto deck can be pumped into slop tanks (other containers) by means of Wilden pumps located at the break of accommodation port and stbd.

5. Five persons will be on duty during any cargo transfer operation:

Person in Charge :	Chief Officer
Asst. in Charge :	Duty Officer and pumpman
Deckhands :	2 persons (bosun, Abs, OS and TNO)

6. One deck hand with walkie talkie will always be at cargo manifold, the other one to be attending to moorings and antipollution watch. pumpman's duty will be directed by Duty Officer or Chief Officer. Additional crew will be called and used as conditions demand.

7. In any emergency, cargo operations to be stopped immediately. Cargo pumps can be tripped using emergency stops at cargo control room, both sides of manifolds, Forecastle and Poop deck areas.

9. Communication is by 8 UHF walkie talkie sets with key persons.

	CHEMICAL TANKER MANUAL	Prepared by	Marine Supdt
		Checked by	Director
		Edition	Edition No.4
		Issue date	Mar 24
		Approved by	Director
		Rev date	-
		Page	3 of 8

10. Cargo tanks will be topped up using Hermetic Portable Cargo Monitoring Equipment / with remote readout at the cargo control room. Gauges' accuracy to be verified when ullage is 2.5 metres. After topping up a tank, watch that the ullage is steady and confirm no more cargo coming in. All tanks are also fitted with independent high-high level alarms, these are float sensing type (98% per cargo tank). This gives an audible and visual alarm on the main deck (fitted on bridge-front) and in the Cargo Control Room.

11. Once transfer operations are complete, Chief Officer will personally check that all valves opened during the operation are now shut.

12. In the event of any oil spill or discharge, sound the 'Emergency Alarm' stop cargo operations and inform Master, also advise terminal. Local USCG Group is to be informed on VHF Channel 16 of situation. It is most important to contain any spill on deck. Do not spray any chemical in water unless USCG or Port Authority gives permission. Be familiar with ship's oil spill containment equipment and location. Proceed as per Vessel Response Plan/SOPEP.

App 1.8 DESCRIPTION OF THE CARGO SYSTEM, OPERATIONAL AND EQUIPMENT PARAMETERS US CFR 33 155.750 (2) (i) (ii)

Reference ship's drawing no.5105da800d101, diagrammatic Arrangement of cargo oil and water ballast system (note: copy of this dwg is conspicuously posted in cargo Operations control room on accommodation a-deck).

Main features of the oil transfer system

The Cargo system in this vessel consists of a total of 20 cargo tanks and 2 dedicated slop tanks. The cargo tanks are arranged as 10 pairs of wing tanks. The vessel is a double hull vessel with six pairs of wing tanks, 3 DB Centre and Fore Peak and Aft Peak arranged as segregated ballast tanks. This satisfies the regulations of the MARPOL 73/78 Convention regarding segregated ballast with protective location. The two slop tanks are arranged aft of cargo tanks.

Stripping system

Auto unloading system

Unloading system. Each cargo tanks including slop tanks arranged by separate FRAMO pump and can be used separately or in parallel.

Cargo stripping. The vessel has a separate stripping line from every tank and stripping of cargo following according general FRAMO instruction

Vapour emission control system (VECS)

Vessel is fitted with vapour emission control system. Four vapour connections in the cargo manifold area are installed and painted according to OCIMF standards. The VECS has the following features:

1. Four sets of principal reducers size 600 x 400 and two sets of reserve reducers, size 600 x 300mm. This line is connected to the IG main line system, branching to each cargo tank and to a gas riser.

2. An OVS-OP 30 sampling system, analysing waste vapour gas for oxygen % and pressure. The alarm unit gives the following alarms:

   a. Oxygen High

   b. Flow fail

   c. Vapour Pressure High

	CHEMICAL TANKER MANUAL	Prepared by	Marine Supdt
		Checked by	Director
		Edition	Edition No.4
		Issue date	Mar 24
		Approved by	Director
		Rev date	-
		Page	4 of 8

d. Vapour Pressure Low

The vapour emission control system has two functions:

3. Monitoring of vapour emission from cargo tanks during loading when vapours are led to shore installation. The Omicron independent high alarm system gives warning of oil overflow from a tank getting into IG line, thus avoiding any risk of cargo going into the shore reception facility along with the vapours.

4. The OVS-OP 30 monitors the vapours being received during lightering operations from daughter vessel, if required. Thus, a control over the quality of vapour received can be maintained.

Valve remote control system

The vessel is fitted with an Electro hydraulically controlled Valve Remote Control system (VRC), to operate Cargo and Ballast valves from the CCR Pressurised oil generated from the Power Pack is directed into the opening or closing chambers of the jack, actuating the valves through solenoid operated control valves. The system comprises the following:

1. Hydraulic power pack
2. Solenoid valve rack (installed in safe area)
3. Upper deck boxes
4. Control console
5. Monitored safety box valves
6. Emergency hand pump.

The hydraulic power pack can be selected for constant running only, where the selected motor pump is constantly running. Its pressure and flow compensator control automatically adjusts pump delivery pressure between 0 and 25 ltr per minute, to maintain volume requirements of the system, at the pre-selected operating pressure of 150 bar. The maximum pressure in the circuit is limited to 168 bar, by the relief valve. The following alarms are fitted:

\ High pressure (160 bar)

\ High Temperature (70°C)

An alarm sounds, and the automatic motor pump changes in case of:

\ Low pressure alarm (140 bars).

\ Overload of selected motor pump.

An alarm sounds, and the automatic motor pump set stops, in case of:

\ Low level alarm (the timer allows up to 1 minute operation after alarm).

\ Overload on both motors.

The pump allows one opening/closing of three largest valves simultaneously in 65 secs.

Emergency operation

In case of lack of voltage, operation can be performed from the emergency manual control on the solenoid valves. In case of lack of Hydraulic pressure, by connecting the Emergency hand pump at the following:

CHEMICAL TANKER MANUAL	Prepared by	Marine Supdt	
	Checked by	Director	
	Edition	Edition No.4	
	Issue date	Mar 24	
	Approved by	Director	
	Rev date	-	
	Page	5 of	8

\ On the provided block in the solenoid valve rack or boxes and acting on the emergency manual override of the solenoid valve. The main delivery stop valve should be closed

\ On the block of the safety boxes for submerged valves (main delivery valve closed)

\ On actuator sub plate (for non-submerged valves).

Cargo tank venting system

Each cargo tank is fitted with a PresVac Hi-Lift Hi-Velocity Pressure/Vacuum Valve set for lifting at positive pressure of 1400 mmAq and vacuum pressure of -350 mmAq. These approved P/V valves are provided internally with Flame screens complying with above regulations. Each Cargo tank and SBT is fitted with an Oil Tight Hatch; these hatches are provided with solid NBR sealing seats and must be kept closed at all times. Further, Tank clean ports, steam heating line penetrations, Vapour control valve penetrations, valve remote system hydraulic fluid pipeline penetrations, temperature sensors and level gauge sensor cable conduct penetrations are to be maintained gas tight by periodic visual inspection and soap solution test.

Vessel underway or at anchor:

1. Oil tight hatches on all tanks to be kept closed.

These can be opened only under express instructions of Chief Officer. If vessel is loaded or tanks gassed, then express permission of the Master has to be sought.

2. All cargo system valves connected with the IG system to be kept closed and lashed; these valves can be opened only when necessity of purging via cargo line exists and to be operated only under the instructions of the Chief Officer.

ALL OTHER OPENINGS THROUGH WHICH HYDROCARBON VAPOURS OR LIQUIDS MAY BE RELEASED MUST BE KEPT CLOSED AT ALL TIMES! THIS VESSEL IS FULLY EQUIPPED and CAPABLE OF CONDUCTING ANY OIL TRANSFER OPERATION UNDER FULLY HERMETICALLY CLOSED CONDITIONS.

App 1.9 CAPACITY AND METHODS OF DRAINAGE OF DRIP TRAYS

1. Upon completion of loading/discharging, the contents of loading arms are dropped directly into 4ws or into the drip trays located at the manifolds. This is done by opening the necessary small diameter discharge line valve for each main line (i.e. MARPOL line valve located forward of each manifold valve) and opening the hydrant valve to the drip tray. The capacity of each of the manifold drip trays is 11.60 M3 = 73 bbls. The contents of the drip tray can thereafter be drained into cargo tanks No.4 Wings by gravity by opening the manual valve to the tank. "U" tubes are fitted to these drain lines, inside 4ws COTs, to prevent backflow in case of high IG pressure in tanks. In the event of water collecting in the drip tray, contents can be drained onto deck under supervision of Duty Officer) by opening manual valve located aft of drip tray.

2. While bunkering amidships, contents of arms can be dropped into drip trays of manifold and drained into 4w tanks as described above. If bunkering is to be done aft (forward of accommodation), drip trays are provided each side capacity 2.33 M3 = 14.70 Bbls. Contents of the drip tray can be pumped into any cargo/bunker tank using Wilden pump which is readily available during cargo/bunker operations located strategically in pump room. Water collection in drip tray can be drained onto deck (under supervision) by opening of manual drain plugs provided.

App 1.10 BALLAST SYSTEM

The vessel complies with the regulations for segregated ballast with protective location of the ballast tanks and carries segregated ballast in the Upper Fore Peak tank, Lower Fore Peak tank and in six

	CHEMICAL TANKER MANUAL	Prepared by	Marine Supdt	
		Checked by	Director	
		Edition	Edition No.4	
		Issue date	Mar 24	
		Approved by	Director	
		Rev date	-	
		Page	6	of 8

pairs of wings tanks. The ballast tanks are served by two centrifugal, dual speed electrically driven ballast pumps each having a capacity of 2000/1500 m3/hr. When ballast comes down to stripping level the pumps to be changed over from high to low speed in order to get better stripping results. In order to do this the pump has to be stopped and then restarted at low speed. A maximum of three starts can be attempted in one min. After this, the next start can be attempted only after an interval of twenty minutes. Segregated ballast can also be taken in the Aft Peak tank. This is served by a pump in the engine room having a capacity of 400 m^3/hr. Emergency deballasting can be achieved via a connection with a N/R valve and spool piece arrangement between the cargo and ballast systems in Pumproom. The segregated ballast tanks have a total capacity of 54,559.90 m^3, giving the vessel a ballast draft which satisfies the requirements stipulated in Annex 1 of MARPOL 73/78. Ballast water is not normally allowed to be carried in the cargo tanks, except if this should become necessary under extreme weather conditions, as permitted by Regulation 13(3) of annex of MARPOL 73/78. Such ballast water must then be treated as oily water and the discharge thereof be handled by the cargo pumping and piping system and be monitored by the oil discharge monitoring and control system. Cargo tanks 4W are designated as the normal heavy weather ballast tanks.

THE MANIFOLD VALVES AND SEA SUCTION VALVES ARE TOTALLY MANUAL WHILST THE OVERBOARD VALVE ON THE CARGO SYSTEM IS CONTROLLED BY THE OIL DISCHARGE MONITORING and CONTROL SYSTEM. HOWEVER, THIS VALVE IS USUALLY TO BE KEPT CLOSED AND A BLANK IS FITTED ON THE OUTBOARD SIDE OF THIS VALVE.

SBT SYSTEM SEA VALVES ARE ALL MANUALLY OPERATED, IN PUMPROOM. OPERATION OF ANY SEA VALVE WILL BE UNDERTAKEN SOLELY BY THE CHIEF OFFICER UNLESS IT IS REQUIRED TO SHUT SAME IN AN EMERGENCY AT WHICH TIME CHIEF OFFICER TO BE INFORMED IMMEDIATELY!

OPERATION OF MANIFOLD VALVES SHALL ALSO BE UNDER THE EXPRESS INSTRUCTIONS OF THE CHIEF OFFICER UNLESS DEEMED NECESSARY TO SHUT SAME IN AN EMERGENCY.

BALLAST TANK VENTING SYSTEM AND GAS DETECTION SYSTEM

The vessels ballast line system has a connecting elbow piece with the I.G. piping system. This spool piece can be connected so as to use the I.G. fans in fresh air mode for ventilating the segregated ballast tanks through the bottom ballast lines. The longitudinal Centre line bulk head dividing the Port and Stbd Ballast tanks has manholes which open alternately from port and stbd. W B Tks. These manholes assist in thorough ventilation of the DB tanks. There are four manholes per tank. The Ballast tanks, cofferdam and pump room void spaces are equipped with a gas sampling system (Consilium). This will continuously monitor the atmosphere in all the aforementioned spaces. Before taking ballast, the sampling of the ballast tanks should be switched off and the cocks on the pipelines are to be shut.

App 1.11 Complete ship/shore check list and other relevant checklists prior commencement of cargo operation.

SHIP/SHORE CHECKLIST

Use Form D23.

App 1.12 BALLASTING CARGO TANKS

No. 4W tanks are designated for taking in heavy weather ballast. The ballasting operation should be discussed and agreed in writing between ship, the charterers and terminal. The cargo tanks in which ballast has to be taken must be crude oil washed. Emergency jumper connection between ballast and cargo system to be fitted in Pumproom (painted red colour) OR Sea Suction blank on cargo sea line to be removed. Tank valves for 4P and 4S COT to be opened. Cargo pumps to be started prior opening of sea valves, with vacuum pump running. Pumps are to be slowed down sufficiently in advance of tanks being filled, in order to avoid overflow from tanks being ballasted. Ample ullage space to be left in tanks, even if this means having to top them to normal ullage after departure. When ballasting in non-gas free cargo tanks, the gas must be vented through vent riser.

App 1.13 DISCHARGE OF DIRTY BALLAST

This vessel is an SBT tanker, however if ballast is taken in cargo tanks on account of heavy weather or the charterer's requirements, in such cases, discharge of dirty ballast must always be performed according to Regulation 9 of Annex I Chapter II of MARPOL 73/78.

Quote:

"Discharge of oil or oily mixtures from ships shall be prohibited except when all the following conditions are satisfied:

A. Tanker is not within a special area.

B. Tanker is more than 50 nautical miles from nearest land.

C. Tanker is proceeding en route.

D. Discharge of oil content does not exceed 30 litres per nautical mile.

E. Total quantity of oil discharged into the sea does not exceed 1/30000 of total quantity of particular cargo.

F. ODME should be operational."

Unquote

Note: Discharge of "dirty" ballast is prohibited in special areas.

Procedures

Discharge of dirty ballast at sea must always be performed under strict control of pumping procedures.

1. Before discharging the dirty ballast overboard, flush the main cargo lines to be used for discharging dirty ballast into the slop tank.

2. Before flushing, prime the system, establish suction, stop the pump, close all valves and allow the oil from the pipe walls to separate out.

3. Resume pumping after half an hour at moderate rate.

App 1.14 DISCHARGING DIRTY BALLAST

1. Commence discharge dirty ballast.

2. Reduce discharge rate from individual tanks on approaching a water depth of about 20% of the tank depth.

3. Reduce pumping rate.

4. Observe carefully trend of oil monitor reading.

5. Stop discharge of individual tank when a level is reached that is known not to give any entrainment of oil.

6. Transfer remaining dirty ballast into the slop tank using the stripping system.

7. After sufficient settling time (heating improves settling) the stripping pump is to be used to decant slop tanks.

CHEMICAL TANKER MANUAL

Prepared by	Marine Supdt		
Checked by	Director		
Edition	Edition No.4		
Issue date	Mar 24		
Approved by	Director		
Rev date	-		
Page	8	of	8

Engine room bilges

Vessel can only transfer oil from the bilge tank can only be transferred to slop tank, if mentioned in the IOPP Form B.

App 1.15 DECANTING SLOP TANKS

Decanting of slop tanks is a critical step in the retention of oil on board. Even a short delay in stopping a pump or closing a valve can allow oil to escape into the sea. Time required for oil and water to separate in the slop tank depends on the motion of the ship, temperature and type of cargo. Under favourable conditions a few hours may be enough but in most circumstances up to 36 hrs should be allowed. Discharge from the slop tank must cease well before the interface is reached. Carefully the trend of the oil monitor reading. The pumping rate is to be reduced to a minimum well in time, before the oil/water interface is reached. If oil should appear before the predetermined ullage is reached, stop pumping immediately.

END OF PROCEDURE

CHECKLIST NO. 1: PRIOR DISCHARGE

	CHEMICAL TANKER MANUAL	Prepared by	Marine Supdt		
		Checked by	Director		
		Edition	Edition No.4		
		Issue date	Mar 24		
		Approved by	Director		
		Rev date	-		
		Page	1	of	1

Procedure document 2
Checklist No.1 – Prior discharge
Dated 20240311

VESSEL NAME	
PORT	**DATE**

		YES	NO	N/A
1	Cargo Discharge Plan (including Ballast Operations) ready and confirmed by Master?			
2	Oil Record book (Part 2) up to date?			
3	All valves in the cargo system are shut prior to lining up?			
4	Ballast Management Log up to date?			
5	Officers and Crew briefed?			
6	Is the vessel with acceptable stability and stress?			
7	Portable gauging equipment ready for use?			
8	Oxygen meters calibrated and ready for use?			
9	Walkie-talkies ready for use?			
10	Safety plan at Manifold and warning board displayed?			
11	Is the firefighting equipment rigged and ready for immediately use?			
12	Pilot ladder, gangway, life ring w/light checked and ready?			
13	Midship Cargo Hose Cranes Tested for satisfactory operation & overload Trips			
14	All external doors, vents, E.R. skylight closed, A.C. on partial recirculation?			
15	Garbage drums on deck covered and Garbage Record book up to date?			
16	Radio and radar transmitters switched off and equipment grounded?			
17	Vessel earthed?			
18	Deck scuppers plugged? Are all drains on both Spill Tanks fully closed? Tanks empty?			
19	Air-driven pumps on both side aft Cargo deck and Oil spill containers ready?			
20	Cargo & Ballast pumps tested and ready for use?			
21	Check setting of pump temperature trips and alarms is based on the flash point of the cargo.			
22	COP and I.G. gauges (Both remote and local) are checked and at manifold where gauges not fitted, plugs in position. Remove covers from the manifold pressure gauges.			
23	Independent High – High level alarms to try out and check in order.			
24	Cargo Remote level gauging high and low level alarms to check is order			
25	The cargo tank pressure sensors are set correctly and checked for satisfactory operation (If fitted)			
26	SBT tanks checked for oil. Sounding and L.E.L. readings taken.			
27	Is the cargo lines Pressure Test not expired?			
28	Lower IG pressure for gauging/river transit to below 200 mm (if applicable)			
29	Hydraulic System for Cargo valves operations, oil level checked?			
30	All Manifold connections checked by OOD and ready for Operations?			
31	Manifold area, Ballast Room atmosphere Gas detector ready?			
32	Are P/V valves (or Vapor Return line) ready for Operations?			
33	Is the Ship/Shore safety Check list signed prior discharging?			
34	Is the Ship/Shore Discharge Agreement signed prior discharging?			
35	Is AIS and VHF on low power?			
36	Is Manifold clear of all non-metallic containers and pans which are not grounded?			
37	Are all unused manifold valves shut and piping blanked with pressure gauge in full operational view for check during security and safety rounds?			
38	Pump Room Bilges dry/clean. All Pumproom lights/shades checked. One flashlight kept at pumproom bottom.			
39	Pumproom gas alarms and Bilge alarm checked.			
40	Pumproom entry checklist filled?			
41	Is the pumproom bilge system set so that it is operational from pumproom top?			

CHIEF OFFICER	**BOSUN**	**CHIEF ENGINEER**
2ND OFFICER	**MASTER**	**ELECTRICIAN**
3RD OFFICER		**PUMPMAN**

CHECKLIST NO.2: PRIOR LOADING

	CHEMICAL TANKER MANUAL	Prepared by	Marine Supdt
		Checked by	Director
		Edition	Edition No.4
		Issue date	Mar 24
		Approved by	Director
		Rev date	-
		Page	1 of 1

Procedure document 3
Checklist No.2 – Prior loading
Dated 20240311

VESSEL NAME	
PORT	**DATE**

		YES	NO	N/A
1	Cargo Discharge Plan (including Ballast Operations) ready and confirmed by Master?			
2	Oil Record book up to date?			
3	Ballast Management Log up to date?			
4	Officers and Crew briefed?			
5	Is the vessel with acceptable stability and stress?			
6	Portable gauging equipment ready for use?			
7	Oxygen meters calibrated and ready for use?			
8	Walkie-talkies ready for use?			
9	Safety plan at Manifold and warning board displayed?			
10	Is the firefighting equipment rigged and ready for immediately use?			
11	Pilot ladder, gangway, life ring w/light checked and ready?			
12	Midship Cargo Hose Cranes Tested for satisfactory operation & overload Trips			
13	All external doors, vents, E.R. skylight closed, A.C. on partial recirculation?			
14	Garbage drums on deck covered and Garbage Record book up to date?			
15	Radio and radar transmitters switched off and equipment grounded?			
19	Vessel earthed?			
20	Deck scuppers plugged?			
21	Are all drains on both Spill Tanks fully closed? Tanks empty?			
22	Air-driven pumps on both side aft Cargo deck and Oil spill containers ready?			
23	Ballast pumps tested and ready for use?			
24	Independent High – High level alarms to try out and check in order.			
25	Cargo Remote level gauging high and low level alarms to check is order			
26	Hydraulic System for Cargo valves operations, oil level checked?			
27	All Manifold connections checked by OOD and ready for Operations?			
28	Manifold area, Ballast Room atmosphere Gas detector ready?			
29	Are P/V valves (or Vapor Return line) ready for Operations?			
30	For Pressure sensors provided for COT's as means of secondary protection are the alarm settings correct? **Note:** The pressure sensors alarms should be set to activate.			
31	Is the Ship/Shore safety Check list signed prior loading?			
32	Is the Ship/Shore Discharge Agreement signed prior loading?			
33	Is the Cargo Safety Data Sheet delivered by Loading Master prior loading?			
34	Is AIS and VHF on low power?			
35	Is Manifold clear of all non-metallic containers and pans which are not grounded?			
36	Are all unused manifold valves shut and piping blanked with pressure gauge in full operational view for check during security and safety rounds?			
37	Pump Room Bilges dry/clean. All Pumproom lights/shades checked. One flashlight kept at pumproom bottom.			
38	Pumproom gas alarms and Bilge alarm checked.			
39	Pumproom entry checklist filled?			
40	Is the pumproom bilge system set so that it is operational from pumproom top?			

CHIEF OFFICER	**BOSUN**	**CHIEF ENGINEER**
2ND OFFICER	**MASTER**	**ELECTRICIAN**
3RD OFFICER		**PUMPMAN**

CHECKLIST NO.3: DEPARTURE PORT

	CHEMICAL TANKER MANUAL	Prepared by	Marine Supdt
		Checked by	Director
		Edition	Edition No.4
		Issue date	Mar 24
		Approved by	Director
		Rev date	-
		Page	1 of 1

Procedure document 4
Checklist No.3 – Departure port
Dated 20240311

VESSEL NAME	
PORT	**DATE**

		YES	NO	N/A
1	All crewmembers on board and accounted for.			
2	FFA equipment / special equipment restowed for sea.			
3	Vessel checked for stowaways - Entry in logbook.			
4	Oil spill equipment restowed for sea .			
5	Anchor lashings taken. Cement on spurling pipe on loaded voyage. Hydraulic power switched off.			
6	Heavy weather lashings taken , if weather warrants all rooms/stores on deck properly closed.			
7	Pilot access ladder / combination , secured for sea			
8	Accommodation ladders secured for sea			
9	Moorings systems – power switched off / Mooring ropes covered with canvas			
10	Deck scuppers to be opened once clear of Port limits .			
11	All loose gear restowed / secured for sea			
12	Sounding pipe caps checked / closed			
13	Drip trays / save all trays - plugs unshipped			
14	If in heavy ballast top up the specific heavy ballast tanks for the ship			
15	Check Eye showers on deck are secured and operational			
16	Check lashing / securing arrangement of lube oil drums On the deck aft and garbage drums on poop deck .			
17	When leaving a port with known stowaway problems , carry out a thorough stowaway check again after Drop of Pilot			
18	All Cargo lines are well drained and that no shift of cargo takes place			
19	Canvas covers are put on PV valves,Tank domes and Butter worth openings depending on the cargo			
20	Nitrogen if used for padding then Nitrogen pressure is maintained			
21	MMC,Sounding tapes are well cleaned and secured			
22	If vessel equipped with Framo system that Framo pumps are purged			
23	Stability condition for Departure port printed out and filed			
24	All valves in pumproom and deck are secured			
25	Cargo samples secured in designated Sample locker			
26	Hose handling,Provision cranes and Bunker davits hooks properly secured			
27	Draft gauge valves are shut			
28	Loose equipment in Lifeboat secured for sea			
29	If vessel in loaded condition the Ballast tank gas monitoring system is on			

CHIEF OFFICER **BOSUN** **CHIEF ENGINEER**

2ND OFFICER **MASTER** **ELECTRICIAN**

3RD OFFICER **PUMPMAN**

BUNKER TRANSFER PROCEDURES

	CHEMICAL TANKER MANUAL	Prepared by	Marine Supdt
		Checked by	Director
		Edition	Edition No.4
		Issue date	Mar 24
		Approved by	Director
		Rev date	-
		Page	1 of 6

Procedure document 5
Bunker transfer procedures
Dated 20240311

1.0 BUNKERING TRANSFER PROCEDURES

The guidelines noted hereunder refer to any procedure which involves the transfer of fuel oil, diesel oil or lubricating oil to and from ship or from one tank to another and not considered as cargo.

1.1 PERSONNEL AND THEIR DUTIES

1.1.1 OVERALL IN CHARGE - CHIEF ENGINEER

Undertakes overall charge of procedure and is responsible for carrying out the task in proper manner, allocate related duties to other persons on board and oversee their work. Entire operation can be broken down in to following steps.

1.1.1.1 Determination of the amount of oil that the ship can take and order it through the Owners or Charterers. Accurate description and specification of the bunkers must be provided to the company arranging bunkers.

1.1.1.2 Preparation of the bunkering plan.

1.1.1.3 Establishment of the bunkering organisation and allocation of related duties to other staff.

1.1.1.4 Verification of preparations and safety precautions.

1.1.1.5 Liaison with bunker suppliers.

1.1.1.6 Supervision of testing of samples at the start of and during the bunkering operation and overall guidance and control.

1.1.1.7 MSDS to be obtained for each bunker

1.1.2 ASSISTANT TO CHIEF ENGINEER - BUNKER OFFICER (3RD OR 4TH ENG. AS NOMINATED BY C/E).

Undertake the following duties:-

1.1.2.1 Supervision of overall bunker operation including the preparation, bunkering work & post bunkering restoration work giving detailed instruction to workers.

1.1.2.2 Liaison with bunker barges.

1.1.2.3 Taking sounding. Also, bunker barge initial and final soundings/ullage of all tanks and acknowledgement taken of this from barge Master.

1.1.2.4 Checking on oil Specific Gravity & calculation of amount of oil received.

1.1.2.5 Checks on ships heel and trim.

1.1.2.6 Watch keeping on oil spillage.

1.1.2.7 Verification and reporting of the completion of bunkering operation.

1.1.2.8 Make entry in the checklist.

1.1.3 ASSISTANT TO BUNKERING OFFICER (TWO ER RATINGS)

Undertake the following duties under supervision of bunkering officer: -

	CHEMICAL TANKER MANUAL	Prepared by	Marine Supdt
		Checked by	Director
		Edition	Edition No.4
		Issue date	Mar 24
		Approved by	Director
		Rev date	-
		Page	2 of 6

1.1.3.1 Check Air Vent for flow of Air

1.1.3.2 Arrangement of empty trays for receiving leaked oil.

1.1.3.3 Preparation of equipment & materials.

1.1.3.4 Open/Close operation of valves under instruction only.

1.1.3.5 Maintain watch on oil spill/leaks.

1.1.3.6 Pre and Post bunkering restoration work.

1.1.4 SPARE MEN - AS DESIGNATED BY CHIEF ENGINEER

1.1.4.1 As situation demands. Assisting in preparation for bunkering operation, post bunkering restoration work.

1.1.4.2 To provide back up assistance in emergency situation.

1.1.5 DECK DEPARTMENT - PERSONNEL SHOULD ENGAGE IN THE FOLLOWING ITEMS OF WORK TO ASSIST IN THE BUNKERING OPERATION.

1.1.5.1 Mooring/Unmooring operation of oil supply barges.

1.1.5.2 Checking the mooring condition of the ship during bunker operations.

1.1.5.3 Preparation of fenders and pilot ladders etc.

1.1.5.4 Verification of ships draft, trim and heel.

1.1.5.5 Closing and opening of deck scuppers.

1.1.5.6 Arrangements for flying/posting signal flags or lights.

1.1.5.7 Watch keeping for possible spillage accidents on board the ship as well as on the surface in the vicinity of the ship.

1.2 INFORMATION

Following items to be displayed prominently near the transfer facility and must be in large bold letters.

1.2.1 Ships bunkering pipeline diagram to be displayed.

1.2.2 Oil tank layout with associated vent piping.

1.2.3 Tank capacities full/98%/95%/90%

1.2.4 Product description: GENERIC NAME

FUEL OIL - PHYSICAL PROPERTIES:

Viscosity :

Sp. Gravity :

Flash Point :

Colour : Black

Odour : Strong Aromatic

DIESEL OIL - : Details as above.

CHEMICAL TANKER MANUAL

Prepared by	Marine Supdt
Checked by	Director
Edition	Edition No.4
Issue date	Mar 24
Approved by	Director
Rev date	-
Page	3 of 6

LUBRICATING OIL: Details as above.

1.2.5 PRODUCT HANDLING PRECAUTION

The products are non-toxic. In very rare cases can cause allergic reaction to skin. Avoid body contact. Wear impervious gloves. If oil gets into eyes, wash with fresh water for 15 minutes and seek medical help. REFER TO MATERIAL SAFETY DATA SHEET (MSDS)

1.2.6. PRESSURE TESTING OF BUNKER PIPE LINES

Bunker line to be pressure tested to its working pressure annually onboard for vessels less than 20 years old. For vessels more than 20 years old, bunker lines are to be pressure tested to its working pressure once in six months on board. Twice in 5 years, in Dry Dock, it should be tested to 1.5 times of the design pressure. The working pressure should not be mixed up with the design pressure mentioned in the piping diagram booklet. Initially, the working pressure maybe established by checking last 10 or 15 bunker plan where in rate and pressure at which bunker is taken is mentioned. The average higher pressure from this plan should be taken. This will be the established working pressure which should be stenciled on the bunker line on deck near the manifold valve. Annual testing onboard should be done to this pressure. The pressure testing of the line should be done by the F.O. transfer pump. In some cases, the pump relief valve may have to be adjusted to a higher value to be able to achieve the testing pressure on deck. In such event, the setting should adjusted back to its normal value after the test. Before setting to higher pressure, the pump specification should be checked so that relief valve is not adjusted above that value. In case, the transfer pump is not capable of delivering up to the test pressure. Bunker line should be filled up by the F.O transfer pump first and then a hand pump or similar such pump should be used to build up the pressure. Testing pressure and date should be stenciled on the bunker line.

1.3 PRECAUTIONS

Utmost care and vigilance must be exercised while carrying out the bunkering or transfer operations. Following guidelines most be strictly adhered to during operations to avoid any spillage.

1.3.1 Prepare the oil receiving plan in advance taking into account the trim/list are accurately accounted for in calculation of quantities in various tanks. Refer to sample in Section 5.7 of this manual

1.3.2 Bunkering Checklist and the Bunkering Safety Checklist in ship type specific manual is to be complied with and completed. Any additional ship specific item can be included by hand in the bunkering checklist in the ship type specific manual.

1.3.3.1 Bunkering Safety checklist to be signed both by C/Eng and bunker barge/terminal in-charge. The Safety checklist to be used for loading bunkers from barge, when taking bunkers from Jetty or when loading bulk lubricating oil or gas oil from road tanker and is to use in all such cases.

1.3.3.2 All tanks on barge to be sounded and / or flowmeter noted.

1.3.3.3 Proper signal/lights for bunkering posted.

1.3.3.4 Line for bunkering to be set and double-checked. Final soundings agree and bunker team advised. (DB Tanks to not exceed 90% capacity). Unauthorised personnel not to touch valves.

1.3.3.5 Valves in system not in use to be kept shut. Other filling connections to be kept blanked.

1.3.3.6 Ensure O/flow tanks are empty prior beginning operation. Special attention must be paid in this regard in vessels fitted with the ring main/pressure relief systems. O/flow tanks to be sounded at regular intervals and action taken accordingly.

1.3.3.7 Start transfer slowly and once pipe lines attain the temperature of supplied oil, increase rate to the agreed value. Watch the pressure gauge at the bunker manifold closely.

	CHEMICAL TANKER MANUAL	Prepared by	Marine Supdt		
		Checked by	Director		
		Edition	Edition No.4		
		Issue date	Mar 24		
		Approved by	Director		
		Rev date	-		
		Page	4	of	6

1.3.3.8 Maintain record of the filling rate at all times.

1.3.3.9 Ensure sounding in tanks not being filled are remaining same.

1.3.3.10 Check sample for correct viscosity, density and flash point. Any abnormal values to be reported/protested to safeguard interest.

1.3.3.11 Exercise caution when changing over tanks. Open fully the valve of the tank to be filled then close the valve of the filled tank slowly.

1.3.3.12 Check the line pressure at all times against system over pressure.

1.3.3.13 Reduce rate to safe amount while topping up. Top up one tank at a time.

1.3.3.14 Adjust mooring lines during operations.

1.3.3.15 Check ships trim and list condition and maintain in a steady condition.

1.3.3.16 Maintain watch on board for oil spill /leaks.

1.3.3.17 End of transfer:-

(1) Exercise caution when blowing through lines especially if the tank are full especially if the tanks are full to prevent oil spilling from vents.

(2) Secure all valves prior to disconnecting hoses

(3) Fit necessary blank flanges

(4) Check final sounding

(5) Clean up containment trays and put back all materials in store.

1.3.3.18 In case of doubt/difficulty, cease transfer till problem resolved and it is safe to continue.

1.4 POLLUTION CONTROL

In case of accidental spillage and pollution:

1.4.1 Inform authorities concerned without delay

1.4.2 Endeavour utmost to minimise spill overboard

1.4.3 Assist in the cleanup operation

1.4.4 No chemicals to be sprayed on water unless authorised by the local authorities.

1.5 QUALITY AND QUANTITY CONTROL

1.5.1 It has happened that the specific gravity of bunkers being supplied are marginally less than the figures given in the delivery note. The difference may not appear to be large but when the weight of bunker is calculated by using two different Sp. gravities the difference can be significant. Since cost of bunker is based on the weight the price differential can be quite large. Therefore care must be given to checking correct sp. gravity and if difference is noted then it must be brought to the attention of the supplier.

1.5.2 In case of such disputes the bunker sample is to be analysed by an independent laboratory. For this purpose proper samples must be taken from the bunker line in presence of the supplier and sealed with his signature.

	CHEMICAL TANKER MANUAL	Prepared by	Marine Supdt		
		Checked by	Director		
		Edition	Edition No.4		
		Issue date	Mar 24		
		Approved by	Director		
		Rev date	-		
		Page	5	of	6

1.5.3 Importance of viscosity is well known. Sample taken from the line should be immediately analysed to check if viscosity is same as ordered. Small difference, say about 20 CST, can be acceptable. But if the difference is large then it cannot be ignored and must be brought to the attention of the supplier.

1.5.4 There have been incidents when the difference was found to be quite large and the ship had stopped bunkering. The supplier then admitted their mistake and changed the supply. However, in such situation pleased ensure that your testing equipment and procedure are absolutely trustworthy.

1.5.5 With the impact of environmental legislation and awareness, the nature of MDO or HFO contamination is changing. Recently a vessel bunkering from a barge received HFO with a Flash Point dangerously below the 60°C minimum required by Class and SOLAS Chapter II-2, Reg. 15. If there is any doubt on this account it would be prudent to ensure proper custody transfer sampling is done, sent for analysis and bunker not used until analysis received.

1.5.6 Where available tank space makes it possible, bunkers are not to be mixed (old & new). When bunkers are mixed, a sample with the ratio as in the tank to be sent for analysis. The mixed bunkers are not to be used until analysis is received.

1.5.7 Bunker samples during bunkering is to be taken with Drip sampler only. This is a requirement as per IAPP and vessels are to comply with procedures and requirements of Marpol Annex VI.

1.5.8 Vessels are to be aware of ECA areas where only Low Sulphur fuel and diesel is to be used. Ship specific Procedures as per IAPP for changing over to Low sulphur fuel and diesel to be complied with as per ship specific Procedure which have been prepared after discussions between vessels superintendent and the Vessel.

1.5.9 Vessels to be guided by the following for Maximum % capacity that is to be filled for Bunkers:

(a) For DB Tanks, the DB tank percentage is not to exceed 85%. When "Full Bunkers" are to be taken, then, 90% is acceptable, however with slower pumping rate and due caution.

(b) For shoulder or Deep tank, the recommended fill percentage is 95%. This fill percentage can be increase to 98% on a case-by-case basis in consultation with WALLEM when "Full" bunkers are to be taken.

1.5.10 Tankers are to check the vapour space of the bunker tank prior to, during and after bunkering. If H2s is detected, the bunker tank should be periodically tested during voyage and precautions taken as required for avoiding personal exposure to H2S.

1.6 DURING THE VOYAGE

Bunker tanks sounding must be physically checked by a responsible ship's engineer and records of quantities maintained. This physical check of bunker tank soundings must be made at least twice every week, before arrival port and before and after bunkering. Do not rely on flow meters for calculating quantities of bunkers on board. Soundings of all fuel oil and diesel oil bunker tanks must be taken by a responsible ship's engineer and recorded in the Engine Logbook twice per week whilst at sea, at end of voyage, prior to departure and when in sheltered waters at anchor. Please note that this does NOT refer to Sludge tanks as mentioned in Appendix to IOPP certificate. If the voyage is less than three days then soundings are to be taken at the end of voyage and prior to departure only. The Fuel ROB figures are to be calculated by the Chief Engineer using these soundings and the vessels' calibration tables. Temperature corrections, correct S.G. from bunker receipt and trim and heel corrections must be used in these calculations. These figures are to be presented to the Master for his reference.

	CHEMICAL TANKER MANUAL	Prepared by	Marine Supdt
		Checked by	Director
		Edition	Edition No.4
		Issue date	Mar 24
		Approved by	Director
		Rev date	-
		Page	6 of 6

1.7 BUNKER ANALYSIS REPORT

Upon receipt of bunker analysis report C/E must discuss same with entire Engine Team and issue instructions for separator setting, temperature setting and service/settling tank draining requirement and logged down.

	CHEMICAL TANKER MANUAL	Prepared by	Marine Supdt
		Checked by	Director
		Edition	Edition No.4
		Issue date	Mar 24
		Approved by	Director
		Rev date	-
		Page	1 of 3

BUNKERING SAFETY CHECKLIST

Port ____________________ Date ____________________

Ship ____________________ Barge ____________________

Master ____________________ Master ____________________

1. Bunkers to be Transferred

Grade	Tonnes	Vol. @ Loading Temp	Loading Temp	Max Transfer Rate	Maximum Line Pressure
Fuel oil					
Gas oil/diesel					
Lube Oil in bunk					

2. Bunkers to be Loaded

Tank Nos	Grade	Vol. of Tank @ ______%	Vol. of Oil in Tank before Loading	Available Vol.	Vol. to be loaded	Total Vol. Grades

3. Checks by Barge Prior to Berthing

Bunkering	Ship	Barge	Code	Remarks
1. The barge has obtained the necessary permissions to go alongside receiving ship.				
2. The fenders have been checked, are in good order and there is no possibility of metal to metal contact.			R	
3. Adequate electrical insulating means are in place in the barge-to-ship connection.(34)				
4. All bunker hoses are in good condition and are appropriate for the service intended.(7)				
5. The barge is securely moored.(2)			R	
6. There is a safe means of access between the ship and barge.(1)			R	
7. Effective communications have been established between Responsible Officers.(3)			A R	(VHF/UHF Ch ....) Primary System: Backup System: Emergency Stop Signal:
8. There is an effective watch on board the barge and on the ship receiving bunkers.(22)				

	CHEMICAL TANKER MANUAL	Prepared by	Marine Supdt		
		Checked by	Director		
		Edition	Edition No.4		
		Issue date	Mar 24		
		Approved by	Director		
		Rev date	-		
		Page	2	of	3

9. Fire hoses and fire-fighting equipment on board the barge and ship are ready for immediate use.(5)				
10. All scuppers are effectively plugged. Temporarily removed scupper plugs will be monitored at all times. Drip trays are in position on decks around connections and bunker tank vents. (10) (11)			R	
11. Initial line up has been checked and unused bunker connections are blanked and fully bolted.(13)				
12. The transfer hose is properly rigged and fully bolted and secured to manifolds on ship and barge.(7)				
13. Overboard valves connected to the cargo system, engine room bilges and bunker lines are closed and sealed.(16)				
14. All cargo and bunker tank hatch lids are closed.(15)				
15. Bunker tank contents will be monitored at regular intervals.			A R	at intervals not exceeding…….minutes
16. There is a supply of oil spill clean-up material readily available for immediate use.				
17. The main radio transmitter aerials are earthed and radars are switched off. (42)				
18. Fixed VHF/UHF transceivers and AIS equipment are on the correct power mode or switched off.(40)				
19. Smoking rooms have been identified and smoking restrictions are being observed.(36)			A R	Nominated Smoking Rooms Ship/Tanker: Barge:
20. Naked light regulations are being observed.(37)			R	

4. Checks Prior to Transfer

Bunkering	Ship	Barge	Code	Remarks
1. All external doors and ports in the accommodation are closed.(17)			R	
2. Material Safety Data Sheets (MSDS)for the bunker transfer have been exchanged (26)			R	
3. The hazards associated with toxic substances in the bunkers being handled have been identified and understood.(27)			R	H_2S Content .... Benzene Content ....

	CHEMICAL TANKER MANUAL	Prepared by	Marine Supdt		
		Checked by	Director		
		Edition	Edition No.4		
		Issue date	Mar 24		
		Approved by	Director		
		Rev date	-		
		Page	3	of	3

DECLARATION

We have checked, where appropriate jointly, the items of the Check-List in accordance with the instructions and have satisfied ourselves that the entries we have made are correct to the best of our knowledge.

We have also made arrangements to carry out repetitive checks as necessary and agreed that those items coded 'R' in the Check-List should be re-checked at intervals not exceeding _____ hours.

If, to our knowledge, the status of any item changes, we will immediately inform the other party.

FOR SHIP		FOR BARGE	
Name		**Name**	
Rank		**Rank**	
Signature		**Signature**	
Date		**Date**	
Time		**Time**	

Record for repetitive checks:

Date:			
Time:			
Initials for ship:			
Initials for barge:			

ENGINE ROOM PROCEDURES

	CHEMICAL TANKER MANUAL	Prepared by	Marine Supdt
		Checked by	Director
		Edition	Edition No.4
		Issue date	Mar 24
		Approved by	Director
		Rev date	-
		Page	1 of 6

Procedure document 6
Engine room procedures
Dated 20240311

1.0 GENERAL

Please refer to Sec 11 of SBM I.

All vessels have detailed manufacturer's manual for various machinery in the Engine room. The manuals cover both operation of the machinery and their maintenance. All Permits to work including Enclosed space entry and hot work permit are in the Safety manual and to be used as applicable. When maintenance is being carried out on any equipment, the equipment must be isolated (Lock out & Tag out). Isolation permit in Safety Manual to be completed.

Appendix 3 of this Manual contains the Bunker transfer procedure and the Bunker transfer checklist.

Appendix 4 of this manual contains the Engine room checklists.

Sample of Chief Engineers standing orders and Instruction for calling Chief Engineer is provided in this section. Chief Engineer to draw up his own Standing order and have them signed by other Engineers and the posted. Test routines of Emergency equipment are in the SMMS and are to be carried out as per the planned schedule. All Joining engine room staff to be familiarised with the layout and equipment in the engine room by the outgoing staff or by a officer or crew designated by the Chief Engineer. Blackout recovery procedure would be drawn up specific to the vessel and posted in the ECR.

2. CHIEF ENGINEER'S STANDING ORDERS

(Sample - Each Chief Engineer to prepare his own standing orders)

- All Personnel entering the engine room shall be properly attired in full sleeve boiler suit, safety shoes, helmet and earmuffs/plugs.
- Smoking is only allowed in designated smoking area which for engine room is ECR.
- The machinery space is to be kept clean and tidy at all times. Cotton rags are to be disposed only into the covered drums kept for this purpose. All plastics are to be thrown into drum marked "Plastics only".
- All tools are to be returned to their respective places after work. Good housekeeping practices are to be strictly followed.
- All special tools must be under the custody of 2/E and are to be returned to him after use.
- All personnel are to be familiar with their emergency stations and duties. They must also be familiar with the location and operation of LSA/FFA equipment.
- Safety Equipment under a particular engineer's care must be checked every Saturday and any defects rectified immediately and brought to the notice of Chief Engineer.
- Engine room bilges are to be transferred to bilge holding tank, settled for at least 24hrs before being pumped overboard through 15 ppm oily water separator. The OWS is to be operated only under my supervision.
- No Hot work is to be carried out in the Engine room without obtaining a permit as per safety manual (chapter 19) except in the designated space (engine room workshop). The Chief engineer will authorize the hot work in the designated space (engine room workshop) and state the duration of permit for the day on a prominent notice at the work place. No hot work is allowed in the designated space (Engine room workshop) during bunkering operations.

	CHEMICAL TANKER MANUAL	Prepared by	Marine Supdt		
		Checked by	Director		
		Edition	Edition No.4		
		Issue date	Mar 24		
		Approved by	Director		
		Rev date	-		
		Page	2	of	6

- Various Operational checklists from Appendix 3B and Appendix 4B of this manual are available in ECR for identified critical operations. These checklists are to be filled by duty engineer whenever the particular operation is carried out.
- Jobs completed and spare parts consumed are to be entered in SMMS by individual engineers with their own "User-ID". Appropriate history of the repairs is to also be entered in the SMMS.
- Any abnormality, defect or leakage is to be attended immediately and Chief Engineer informed.
- Duty engineer is to attend all alarms promptly and shall take regular rounds of machinery spaces.
- 2/E is delegated to be in charge of engine room personnel and strict discipline to be maintained at all times. All engineers are to carry out their duties diligently and sincerely.
- Any breakdown of equipment or systems is to be entered in the Defect list section of SMMS with date when the repairs are expected to be completed.
- Risk assessment to be carried out as per guidelines in the Risk assessment section of SBM II.
- MOC (Management of Change for vessel) is to be carried out when any equipment is replaced except for replacement in kind(RIK).
- All personnel are to ensure that high standards of "Safety, Operation & Maintenance" are in practice at all times.
- Presence of the Chief Engineer in Engine room or ECR does not relieve the duty engineer of his watchkeeping/other responsibilities.
- No adjustments to automatic control systems, controllers, actuations etc. to be done without the knowledge and permission of the Chief Engineer.
- For UMS vessels, a thorough round of the machinery spaces as per the UMS checklist is to be completed and bridge to be informed prior putting engine room on UMS mode.

2.1 INSTRUCTIONS FOR CALLING CHIEF ENGINEER

(Sample)

Please inform Chief Engineer by phone in the event of the following:

- When in doubt about any variation in plant parameters.
- When two hours' notice is received prior arrival/departure port/anchorage etc..
- At any time when Bridge requests for change of main engine revolutions.
- Any high or low pressure oil pipe leakage is detected.
- Any sea water pipe rupture or leakage is detected.
- Any abnormal noise or vibration is detected in any machinery.
- Breakdown/changeover of important machinery such as Boilers, Generators, Steering Gear pumps etc..
- Vessel's PA system can be used to call Chief Engineer in case he is not available in the cabin.

3. CONTROL OF HAZARDOUS MATERIAL USED ON BOARD

(1) It is company policy that all chemicals, paints, cleaning materials, etc. are handled and stored according to The International Maritime Dangerous Goods Code. The following are general guidelines but are no substitute for consulting the IMDG and makers product information.

(2) All vessels will maintain a file containing hazard data sheets for various materials in use. MSDS must be maintained for the following:

	CHEMICAL TANKER MANUAL	Prepared by	Marine Supdt
		Checked by	Director
		Edition	Edition No.4
		Issue date	Mar 24
		Approved by	Director
		Rev date	-
		Page	3 of 6

(a) Engine Room chemicals

(b) Engine Room lubes and greases

(c) Paints and Thinners

(d) Domestic cleaning materials

(e) Tank cleaning materials

(f) Miscellaneous

Please refer to SHEQ Manual for asbestos free requirements.

(3) Labelling:

(a) BCS being an ISO 14001 certified company only purchases products from standard manufacturers, as such all their products are properly marked. No chemical or hazardous material will be accepted on board without it being properly marked / labelled.

(b) The labels must be in good condition, clearly readable, with all proper hazardous warnings and markings in place. Any packaging not meeting these requirements should not be accepted on board.

(c) Drums brought on board by shipyards, contractors, or sub-contractors must comply with (B).

(d) On occasions vendors or contractors may supply equivalent tank cleaning chemicals or solvents in used drums. In such situations the drums must not be accepted until they are correctly marked. When ordering chemicals at a port not holding a standard stock, it would be prudent to advise supplier in advance regarding marking of drums.

(e) Empty drums used by ship for storage of liquids must have old labels erased and be remarked. Such drums must be thoroughly cleaned out from previous product to avoid any reaction between incompatible chemicals.

(4) Storage:

These must be stored as per the IMDG regulations but following principles are general company policy.

(a) Inflammable and Toxic materials must under no circumstances be stowed inside the accommodation. This includes large quantities of laboratory methanol, fuel samples, cargo samples, domestic cleaning materials. .

(b) Paint and thinners are to be stowed only in the approved paint locker equipped with fixed fire extinguishing system. Paints or thinners must never be stowed in steering compartment or machinery spaces. At the end of the working day they must be returned to the paint store.

(c) Boiler Chemicals, E.R. Cleaning chemicals, Electrical cleaners.

These are in general a mixture of toxic, corrosive or inflammable material. In some cases the chemicals may react with water or heat to give off toxic fumes and in others they could be are inter-reactive.

(d) Drummed Lubes.

Only very small quantities of ready to use oil should be stowed near machinery.

(e) Miscellaneous:

(i) Read and follow the relevant MSDS sheets for any specific instructions.

(ii) Existing procedures for issue of chemicals are to be further intensified, maintained under strict control, and R.O.B. records maintained.

CHEMICAL TANKER MANUAL				
	Prepared by	Marine Supdt		
	Checked by	Director		
	Edition	Edition No.4		
	Issue date	Mar 24		
	Approved by	Director		
	Rev date	-		
	Page	4	of	6

(iii) Chemicals will only be issued by the Officer responsible for its control. At no time is any crew member allowed to enter the chemical locker un-accompanied.

(iv) If drums are to be cut, then this must be done under direct supervision of a responsible officer, and procedures for 'Hot work" strictly followed.

(iv) Prior to any such work being carried out ensure that empty drums are thoroughly rinsed, and drained, and atmosphere checked.

(v) The plugs / caps are to be removed prior to starting any cutting.

(vi) Empty drums must be cleaned and rinsed prior to being stored, and the drum marked as being cleaned, with the caps open.

(vii) Record of inventory to be kept as per sample below

Vessel : ______________________________

Chemical Locker Inventory and Record for the month of ______________

DESCRIPTION	FROM LAST INVENTORY	DELIVERED THIS MONTH	USED THIS MONTH	R.O.B. STOCK

REMARKS

If any drums were emptied during the month, record of same to be mentioned below:

4. ENGINE ROOM EMERGENCY PROCEDURES

4.1 SHIP SPECIFIC PROCEDURES

Ship specific procedures are to be drawn up and kept in a file in the Engine room. At a minimum, the following emergency procedures are to be included:

1. Measures in case of Black – Out (Black out recovery procedure shall be made and posted in the ECR). All Engineers have to be familiar with the Black-out recovery procedure.

2. Manual start of the Emergency Generator

	CHEMICAL TANKER MANUAL	Prepared by	Marine Supdt
		Checked by	Director
		Edition	Edition No.4
		Issue date	Mar 24
		Approved by	Director
		Rev date	-
		Page	5 of 6

3. Flooding of Engine Room – Vessel to also refer to Checklist 15 of Section V of MCCM.

4. Failure of cooling system for Main Engine and Auxiliary Engines

Any other emergencies can be included as applicable to the Engine room

5. Failure of Purifiers

4.2 MEASURES IN CASE OF FIRE IN ENGINE ROOM

In case Engine Room is manned, following procedure is to be followed by the person who notices the fire:

\ Sound the fire alarm.

\ Inform Bridge/ECR.

Fight the fire locally with available firefighting equipment.

The below mentioned measures are taken when the fire is spread out and beyond control with available firefighting equipment in Engine Room platforms.

\ Stop bunkering operations if in progress.

\ Operate emergency stop switch for F.O. & L.O. pumps .

\ Operate emergency stop switch for E/R fans .

\ Operate quick closing valves as applicable.

\ Emergency generator to be started if it has not already automatically started.

\ Emergency fire pump to be started.

\ Close all flaps; dampers and openings to the Engine Room.

\ Proceed to the EHQ.

In case Engine room was unmanned, following action is to be taken:

\ Duty engineer will first muster at the EHQ.

\ Proceed as directed.

\ Depending on the zone of the fire, entry should be made with full protective gear and charged hose from the most suitable entrance to the Engine Room.

Refer to Checklist 8 of Section V of MCCM.

Checklist 8 also details the procedure for evacuation and total flooding in case the fire is out of control.

4.3 MEASURES IN CASE OF GROUNDING (Specific to Engine room)

Grounding checklist is available as
Checklist 3 and 4 in Section V of MCCM and is to be followed.

In case of grounding, following procedure is to be followed with reference to Engine room:

Main Engine is to be stopped if it has not already tripped. Change over to high sea suction.

Sound all double bottom tanks and cofferdam to assess any rupture of hull. Ensure that there is no changes of E/R bilge levels.

Main Engine to be inspected thoroughly. The following to be checked/tested:

\ Holding down bolts and chocks.

	CHEMICAL TANKER MANUAL	Prepared by	Marine Supdt		
		Checked by	Director		
		Edition	Edition No.4		
		Issue date	Mar 24		
		Approved by	Director		
		Rev date	-		
		Page	6	of	6

\ Crankshaft deflection.

\ Mechanical control gear.

\ Main bearing clearances.

- § Intermediate shaft and its bearing to be checked.
- § Stern tube system for possible leakage.
- § Following to be done only after approval discussion with office
- § Turn engine on turning gear. Observe T/Gear load (Amps) on panels after discussion with office. Check for any abnormal noise or vibration.
- § Inspect Steering Gear. Check rudder movement to ascertain that no leakage of hyd. oil or distortion of rudder stock have taken place.
- § Any attempt to refloat the vessel with the help of Main Engine must be out after ascertaining that no damage to engine, gearings and shafting caused due to grounding.

Points of reference for cargo planning

BCH/IBC Code.
Certificate of Class (tank strength for high density cargoes).
Certificate of Fitness for the Carriage of Dangerous Chemicals in Bulk.
CFR 33 (Parts 124–199).
CHRIS (Chemical and Hazardous Response Information Systems) Guide.
Clean Seas Guide for Oil Tankers.
FOSFA (for Oils, Seeds and Fats).
Guide to Port Entry.
ICS Tanker Safety Guide (Chemical).
IMDG Code.
International Safety Guide for Oil Tankers and Manuals (ISGOTT).
MARPOL 73/78 (latest consolidated version).
MSDS for particular cargo carried.
Prevention of Oil Spillage through Cargo Pumproom Sea Valves.
Procedure and Arrangements Manual (Approved by Class).
Safety in Chemical Tankers.
Safety in Oil Tankers.
Ship to Ship Transfer Guide (Petroleum).
Ship's "VEC System Operating Manual" (Approved by Class).
SMPEP.
SOLAS (latest consolidated version).
Supplement to IMDG Code (Including MFAG and EMS).
Tank Cleaning Guide.
Tank Coating Manufacturer's Compatibility Lists.
USCG Chemical Data Guide for Bulk Shipment by Water.

Glossary

The following are some of the terms and definitions, which may prove helpful in the daily dealings with oil and chemical tankers. This list is compiled from various industry sources. While extensive, it should not be considered a complete list of all vessel and charter party-related terms and definitions.

Accommodation ladder. A term applied to a portable flight of steps suspended over the side of a vessel from a gangway to a point near the water, providing any easy means of access from a small boat. Accommodation ladders are usually supplied with two platforms, one at each end. Sometimes called the "gangway ladder."

Adrift. Floating at random; not fastened by any kind of mooring; at the mercy of winds and currents; loose from normal anchorage. A vessel is said to be adrift when she breaks away from her moorings, wharfs and so on.

AER. Abbreviation of "Annual Efficiency Ratio." This is the ratio of a ship's carbon emissions per actual capacity distance (e.g., DWT × nautical miles sailed). The AER uses the parameters of fuel consumption, distance travelled and design deadweight tonnage. It reflects an index based on the tonnage supply.

Aframax carrier. A medium-sized crude oil tanker of 80,000–120,000 deadweight tonnes. Aframax can transport from 500,000 to 800,000 barrels of crude oil and is also used in lightering. A coated Aframax operating in the refined petroleum products trades may be referred to as an LR2.

Aft, after. Towards the stern or the back of the vessel. Between the stern and the midship section of the vessel.

Afterbody. The section of the vessel aft of amidships.

Agency fee. A fee charged to the ship by the ship's agent, representing payment for services while the ship is in port. Sometimes called an "attendance fee."

Air draft. The distance from the vessel's water line to the uppermost point on the vessel, usually the top of a mast or radar tower. When a vessel has to transit areas where there are overhead obstructions (bridges, power lines, cranes, loading arms etc.) it is vital to know what the air draft (draught) will be at the time of transit. The air draft of a vessel will vary depending on the draft of the vessel and its trim.

Allowed laytime. The number of hours allowed for loading and discharging a cargo as stipulated in a Charter Party.

Aloft. Above the deck.

American Bureau of Shipping (ABS). A Classification Society. Under the provisions of the US Load-Line Acts, it has the authority to assign load lines to vessels registered in the US and other countries.

American Petroleum Institute (API). The American Petroleum Institute, founded in 1919, was the first oil trade association to include all branches of the petroleum industry. The API scale is a scale used to define the specific gravity of a cargo. It is used in the crude trades, with the product trades preferring the metric SG scale.

Amidships ("midships"). The middle portion of a vessel.

API gravity (relative density). Means used by the petroleum industry to express the density of petroleum liquids. API gravity is measured by a hydrometer instrument having a scale graduated in degrees API. The relation between API gravity and relative density (formerly called specific gravity) is API Gravity at 15°C (60°F) = 141.5 131.5\Relative Density 15°C/15°C (60°F/60°F)

Articles of agreement. The document contains all particulars relating to the terms of agreement between the Master of the vessel and her crew. Sometimes called the "ship's articles."

Atlantic Basin. The Atlantic Ocean and the ports and seas surrounding it, including the eastern seaboard of the US and Canada, the US Gulf and Caribbean Sea, the northern and eastern coasts of South America, the North Sea, the Black Sea and the Mediterranean Sea.

Backhaul. A deviation to move cargo on the return leg of a voyage for the purpose of minimising ballast mileage and thereby reducing transportation costs.

Ballast. Seawater taken into a vessel's tanks to increase draft, to change trim or to improve stability. Ballast can be taken in segregated ballast tanks (SBT), located externally to the ship's cargo tanks (double hull arrangement) and in fore and aft peak tanks.

- **Ballast movement.** A voyage or voyage leg made without any paying cargo in a vessel's tanks. To maintain proper stability, trim or draft, seawater is usually carried during such movements.
- **Clean.** Term applied to the seawater used for ballast when it is not contaminated by any oil and is carried in clean tanks.
- **Dirty.** Term applied to the seawater used for ballast when it is contaminated with the remnants or residue left in cargo tanks that previously carried crude oil or heavy persistent refined oils.

- **Permanent.** Ballast carried in ship's tanks that were designed to carry nothing else.
- **Segregated/dedicated.** Ballast kept in tanks segregated from cargo pipes and tanks.

Ballast passage. The "ballast leg" of a voyage is differentiated from the "loaded leg."

Ballast pump. A pump used for filling and emptying the ballast tank.

Ballast tank. The tanks used to carry the vessel's ballast. They may be permanent, dedicated or cargo tanks.

Baltic and International Maritime Council Organisation (BIMCO). In total, around 60% of the world's merchant fleet is a BIMCO member, measured by tonnage (weight of the unloaded ships).

Bareboat charter. A Charter under which a customer pays a fixed daily or monthly rate for a fixed period of time for use of the vessel. The customer pays all costs of operating the vessel, including voyage and vessel expenses. Bareboat charters are usually long-term.

Barge. A general name given to a flatbottomed craft specially adopted for the transportation of bulk cargoes.

Barrel. A volumetric unit of measurement equal to 42 US gallons or 158.99 litres. There are 6.2898 barrels in one cubic metre. Note that whilst oil tankers do not carry oil in barrels (although vessels once did in the 19th century), the term is still used to define the volume.

Beam. The width of a ship at its widest part. Also called its "breadth."

Berth. Dockage space for the vessel. Sleeping quarters. Also slang for having a crew position on the vessel.

Bilge. The lower internal part of the hull where the vertical sides meet the bottom. This term applies to both the inside and the outside of the hull. The internal space can be the lower part of a ship's hold or the engine room and serves as a drainage area where accumulated water can run into and be pumped out from.

Bill of lading (B/L). A B/L is the basic document between a shipper and a carrier and a shipper and consignee. It represents the contract of carriage and defines the terms and conditions of carriage. It is the final receipt from the carrier for the goods shown on it and for the condition of the goods. It describes the nature, quantity and weight of the cargo carried. It is also the document of the title of the goods shown.

Bitts. Cast steel heads serving as posts to which mooring lines and cables are secured on a ship.

Boiler room. Compartment in which the ship's boilers are located.

Boilers. Steam-generating units used aboard ships to provide steam for propulsion or for heating and other auxiliary purposes.

Bonded bunkers. Ship's stores that can be delivered under special arrangements directly from a bonded warehouse to the vessel without payment of customs duties.

Bonded stores. Ship's stores that can be delivered under special arrangements direct from a bonded warehouse to the vessel without payment of customs duties.

Boom. A general name given to a projecting spar or pole that provides an outreach for handling cargo.

Boston beam. Boston Harbour can only accommodate vessels with a maximum beam of 27 metres or 90 feet. Vessels which can comply with this (usually under 40,000 MT DWT) are referred to as "Boston Beamers."

Bow. The forwardmost part of a vessel. This area usually houses gear lockers and is the end where anchors and mooring equipment are located.

BPD. Barrels per day. This is a measure of oil output, represented by the number of barrels of oil produced in a single day.

Breadth See "beam."

Bridge. A general term referring to that area of a vessel where the wheelhouse and chart room are located. It is the navigating section of a vessel.

Bridge aft. Vessels with no midship house. All quarters with the bridge are contained in one superstructure at the after end of the vessel.

Bulbous bow. A large protruding bow section is designed to break water friction allowing the vessel to make better speeds.

Bulk cargo. Bulk cargo is commodity cargo that is transported unpackaged in large quantities. The containment for this type of cargo is the tanks of the ship.

Bulkhead. A partition in a ship that divides the interior space into various compartments in the walls of a vessel's tanks.

Bum boat. A small open rowboat employed in carrying supplies for sale to vessels in a harbour.

Bunkers. Fuel for a vessel. The type will vary depending on the propulsion mode of the vessel. Steamships will use a heavy fuel oil, diesels use a range of fuels from heavy to light and gas turbines use kerosene.

Buoy. A floating object employed as an aid to mariners to mark the navigable limits of channels, their fairways, sunken dangers, isolated rocks, telegraph cables etc.

Butterworth tank cleaning system. A mechanical device used for the purpose of cleaning oil tanks by means of high-pressure jets of hot water. The apparatus consists of double opposed nozzles which rotate slowly about their horizontal and vertical axis and project two streams of water through all angles against all inside surfaces of the space being cleaned. The tank washing machines can deliver sprays of water at various temperatures and pressures that are dictated by the type of cargoes carried and the reasons for cleaning (quick bottom wash through gas-freeing and tank entry for hot work).

Calibration. The process of adjusting or measuring the performance of a device.

Calibration tables (gauge or tank tables, innage/ullage tables). Tables developed by recognised industry methods that represent volumes in each tank according to the liquid (innage) or empty space (ullage) measured in the tank. The tables are entered with linear measurements (for example, feet, inches, metres, centimetres) to obtain calibrated volumes (for example, barrels, cubic metres or cubic feet).

Call letters. The letters are assigned to the ship's radio (station).

Camber. The arching of the deck upward is measured at the centreline in inches per foot beam.

Cancellation clause. A clause in a charter party whereby the charterer reserves the right to cancel the charter if the ship fails to arrive, ready to load, on a specified date at a named port.

Cancelling date. A stated date after which, if a vessel is not ready to load, the intending charterers have the option of cancelling the charter. The passing of the cancelling date leaves the owner's obligation unimpaired unless the charterer releases him.

Capacity. The volume of a container or tank filled to a specified level.

Capacity plan. A general plan or inboard profile which gives all data relating to the capacity of cargo spaces, tanks, bunkers and storerooms.

Capping. Routing a vessel around the Cape of Good Hope, South Africa.

Carbon Disclosure Project (CDP). The Carbon Disclosure Project is a not-for-profit charity that runs the global disclosure system for investors, companies, cities, states and regions to manage their environmental impacts. The world's economy looks to CDP as the gold standard of environmental reporting with the richest and most comprehensive dataset on corporate and city action.

Carbon intensity indicator (CII). The Carbon Intensity Indicator is a response to the company's need to move towards a business model compatible with the Paris Agreement, achieving net zero emissions by 2050. This indicator is used to monitor progress and apply the most suitable and timely efficient levers.

Cargo hose. A hose usually of 15–25 centimetres (6–10 inches) in diameter used for the transfer of cargo from ship to shore and vice versa.

Cargo plan. A plan giving the quantities and description of the various grades carried in the ship's cargo tanks.

Cargo pump. Pump used on tankers for discharging cargo and loading or discharging ballast. Located, at the bottom of the pump room, these pumps are usually of the common duplex type or turbine type of which the centrifugal is the most common.

Cargo quantity option certificate. A certificate signed by vessel and shore representatives acknowledging the amount of cargo intended to be loaded.

Catwalk. A raised bridge running fore and aft from the Midship House and also called the "walkway." It affords safe passage over the pipelines and other deck obstructions.

Centre tanks. Cargo tanks located on the vessel's centreline.

Centreline. A horizontal fore-and-aft reference line, dividing the vessel into two symmetrical halves.

Centrifugal pump. A pump consisting of a shaft to which vanes are attached and which rotates in a circular casing. Water or liquid flows into the casing near the centre of the rotating shaft and is propelled outward along the vanes by centrifugal force. It escapes through a discharge pipe at the circumference of the casing.

Chain locker. The compartment for storing the anchor chains, located near the bow of the ship.

Charter. Contract entered into with a customer for the use of the vessel for a specific voyage at a specific rate per unit of cargo (voyage charter) or for a specific period of time at a specific rate per unit (day or month) of time (time charter).

- **Bareboat charter.** Owner leases an unmanned ship for a long period at a rate that covers any depreciation and nominal return. The charterer mans the vessel and pays all operating expenses.
- **Charter party.** A document of contract, or agreement, by which a shipowner agrees to lease, and a charterer agrees to hire, an entire ship or all or part of the cargo space to carry cargo for an agreed sum under certain conditions.
- **Charter rates.** The tariff applied for chartering tonnage in a particular trade.
- **Charterer.** The company or person given the use of the vessel for the transportation of cargo or passengers for a specified time.
- **Disponent owner.** Charterer who has sublet the vessel and is acting as the owner per the terms of the contract.
- **Period charter.** Refers to consecutive voyage (C/V) exceeding four voyages, time charters (T/C) and bareboat charters.
- **Spot (voyage) charter.** A charter for a particular vessel to move a single cargo between specified loading port(s) and discharge port(s) in the immediate future. Contract rate (spot rate) covers total operating expenses such as port charges, bunkering, crew expenses, insurance, repairs and canal tolls. The charterer will pay all cargo-related costs.
- **Time charter (T/C).** A charter for varying periods of time, usually between two and ten years, under which the owner hires out the vessel to the shipper fully manned, provisioned, stored and insured. The charterer is usually responsible for bunkers, port charges, canal tolls and any crew overtime connected with the cargo. The

charter rate (hire) is quoted in terms of a cost per month per deadweight tonne.

- **To fix a charter.** To reach a final agreement on the terms of a charter party.

Classification of petroleum. Classes "AC" of petroleum is considered flammable and has a flash point of 26°C (80°F) or below. Examples of these classes range from light naphtha (Class A) to most crude oils (Class C). Class D cargoes such as kerosene and heavy crudes are considered combustible and have a flash point above 26°C (80°F) but below 65°C (150°F). Class E cargoes are the heavier fuel oils and lubricating oils and have a flash point above 65°C (150°F).

Classification society (Class). The professional organisations which class and certify the strength and seaworthiness of vessel construction. Class and certification issued to each vessel may be required for insurance purposes. American Bureau of Shipping (ABS) and Lloyds Register of Shipping (LRS) are two of the most well-known classification societies in the world today.

Clean petroleum product (CPP). Liquid products refined from crude oil, whose colour is less than or equal to 2.5 on the National Petroleum Association scale. CPPs include naphtha, jet fuel, gasoline and diesel/gasoil.

Clean service. Tanker transportation of products lighter than residual fuels, for example, distillates, including No. 2 Heating Oil.

Clean ship. Refers to tankers that have their cargo tanks free of traces of dark persistent oils that remain after carrying crudes and heavy fuel oils.

Clingage. The residue that adheres to the inside surface of a container, such as a ship's tank or shore tank, after it has been emptied.

Closed gauging system. A method of obtaining measurements of the tank contents without opening the tank. This may be accomplished by using automatic tank gauges or by taking measurements through a pressure/vapour lock standpipe. This type of gauging is done extensively on vessels with inert gas systems. Such a system that allows no vapours to be lost to the atmosphere is a true closed system while other types that allow minimum vapours to be lost to the atmosphere are called "restricted systems."

Cofferdam. The narrow, empty space between two adjacent watertight or oil-tight compartments. This space is designed to isolate the two compartments from each other and/or provide additional buoyancy. It prevents any liquid contents of one compartment from entering the other in the event of a bulkhead failure. In oil tankers, cargo spaces are usually isolated from the rest of the ship by cofferdams fitted at both ends of the tank body.

Coils. These are the heating elements in the cargo tanks which are used to heat dirty petroleum product (DPP) cargoes on the laden voyage so that the cargo is fluid enough to discharge efficiently at the discharge port. Alternatively, some vessels are fitted with heat exchangers.

Combined carriers or combination carriers. Ships designed to be able to carry both oil cargoes or "dry" cargoes such as iron ore or grain, in the hope that the vessel can maximise earnings by swapping between wet and dry cargoes and eliminate some of the time spent "in ballast." Note that the vessel alternates cargoes and does not carry both oil and dry cargo at the same time.

Commercial management or commercially managed. The management of the employment, or chartering, of a vessel and associated functions, including seeking and negotiating employment for vessels, billing and collecting revenues, issuing voyage instructions, purchasing fuel and appointing port agents.

Consignee. The person to whom cargo is consigned as stated on the bills of lading.

Consignor. The shipper of the cargo.

Contamination. The result from commingling of a grade of cargo with a sufficient quantity of another grade to destroy the characteristics of the cargo.

Contract of affreightment (COA). A Contract of affreightment is an agreement providing for the transportation between specified points for a specific quantity of cargo over a specific time period but without designating specific vessels or voyage schedules. This allows flexibility in scheduling since no vessel designation is required. COAs can either have a fixed rate or a market-related rate.

Controlled fleet. All ships owned and period chartered by affiliate(s).

Cross haul. Two ships on intersecting trade routes. This voyage pattern may indicate an uneconomic vessel location. For example, Aruba/Fawley and Puerto la Cruz/New York.

Crude oil wash (COW). A method of cleaning tanks using oil from the ship's cargo. COW is normally used when a tanker is discharging. Oil is taken from the tanks and pumped through a special line to fixed or semifixed tank washing machines where it is sprayed against all inside surfaces of the tank. This procedure removes any cargo which is "clinging" to the surfaces of the tank.

Crude oil washing. The use of a high-pressure stream of crude oil cargo to dislodge or dissolve clingage and sediment from the bulkheads, bottom and internal tank structures of a vessel during the discharge operation. Note that regulatory agencies require that a vessel's tanks be inerted before this tank cleaning method is used.

CST or centistokes. A measurement scale for the viscosity of a liquid.

Cubic capacity. The inside measurement of a tanker's cargo compartments or tanks, usually expressed in barrels or cubic feet/metres.

Cubic limitation. Reaching cargo tank capacity before the vessel sinks to its load line. This is usually caused by loading light crude (crude with a high API) or clean products.

Davits. A set of arms on a ship from which its lifeboats are suspended.

Dead freight. Non-utilisation of cargo carrying capacity on a vessel.

Deadweight (deadweight tonnage [DWT]). The lifting or carrying capacity of a ship when fully loaded. This measure is expressed in long tons when the ship is in salt water and loaded to her marks. When loaded to her summer marks the value is for her summer deadweight (SWDT). It includes cargo, bunkers, water, (potable, boiler, ballast), stores, passengers and crew.

1 long tonne = 2,204.6 pounds.

Deadweight scale. A table that is part of the vessel plans and indicates the draft the vessel will be down to at any particular phase of loading.

Deck. A platform or horizontal floor that extends from side to side of a ship. The main deck is the highest complete deck on a ship (the one which runs the full length of the ship).

Deck log. Also called the "Captain's Log," "scrap logbook" or "rough logbook." A full nautical record of a ship's voyage, written up at the end of each watch by the deck officer on watch (OOW). The principal entries are course steered; distance run; compass variations, sea and weather conditions; ship's positions, principal headlands passed; names of lookouts and any unusual happenings such as fire, collision and the like.

Deck officer. As distinguished from engineer officer, refers to all officers who assist the Master in navigating the vessel when at sea and supervise the handling of cargo when in port.

Deck stores. The spare gear and consumable stores provided for the upkeep and safe working of the tanker and her cargo, excluding stores used in the engine room.

Deep water route. A designated area within definite limits which has been accurately surveyed for clearance of sea bottom and submerged obstacles to a minimum indicated depth of water.

Demise charter. Also called bareboat charter in which the bare ship is chartered without crew; the charterer, for a stipulated sum, taking over the vessel with a minimum of restrictions usually for 10 or more years. See Bareboat Charter.

Demurrage amounts. Additional revenue paid to the shipowner on its voyage charters for delays experienced in loading and/or unloading cargo that are not deemed to be the responsibility of the shipowner. The revenue is calculated in accordance with specific charter terms.

Density. The density of a homogeneous substance is the ratio of its mass to its volume. The density varies as the temperature changes, and it is usually expressed as the mass per unit volume at a specified temperature.

- **Absolute density.** The mass of a substance per unit volume at a specified temperature.
- **Relative density.** The ratio of the mass of a given volume of fluid to the mass of an equal volume of pure water at the same temperature and pressure. Relative density replaces the term "specific gravity."
- **Relative density at 15°C (60°F).** Fluid relative density measured against water with both materials at 15°C (60°F) and a reference pressure of 14.696 psi (or equilibrium pressure). Equivalent to "RD 15/15" or "RD 60/60".

Deviation. A departure from a voyage pattern on either the forward or return leg of a voyage.

Dirty ballast. Applies to the seawater used for ballast when it is contaminated with the remnants or residue left in cargo tanks that previously carried crude persistent refined oils.

Dirty petroleum product (DPP). Liquid products refined from crude oil, whose colour is greater than 2.5 on the National Petroleum Association scale. DPPs will usually require heating during the voyage as their viscosity or waxiness would make discharge difficult at ambient temperature. DPPs include fuel oil, low sulphur waxy residue (LSWR) and carbon black feedstock (CBFS).

Dirty ship. Refers to tankers that have been carrying crude oil and heavy persistent oils such as fuel oil and dirty diesel oils.

Dispatch. The function of issuing voyage instructions or sailing orders to vessels. Also, an agreed amount to be charged by terminals for prompt vessel turnaround.

Dispatch days. Days saved in the loading and discharge of a cargo vessel within the (lay) time allowed under the charter party. Note that dispatch is not usually applied in the tanker business.

Displacement tonnage. Expressed in tons, it is the weight of the water displaced by the vessel which in turn is the weight of the vessel at that time. The vessel's light displacement is the weight of the vessel only and the vessel's loaded displacement is the weight of the vessel and all cargo, stores, fuel, water etc. onboard.

Disponent owner. Charterer who has sublet the vessel and is acting as the owner per the terms of the contract.

Document of compliance (DOC). This is a certificate issued to the operating company to attest to its compliance with the ISM (International Safety Management) rules.

Double bottom. A general term used for all watertight spaces contained between the outside bottom plating, the tank top and the margin plate. Double bottoms are usually subdivided into a number of separate tanks and can be used to hold clean ballast, potable or boiler-feed water or fuel. They also provide a measure of protection for cargo tanks if bottom plating is damaged in the event of grounding. Chances of pollution may be diminished due to this protection.

Double hull (often referred to as double skin). A design of tanker which has double sides and a double bottom. The spaces created in the double sides and bottom are used for ballast and provide a protective distance between the cargo tanks and the outside world.

Double-sided. Hull construction design in which a tanker has an inner and outer side.

Draft. The vertical distance measured from the lowest point of a ship's hull to the water surface. Draft marks are welded onto the surface of a ship's plating. They are placed forward and aft on both sides of the hull and also amidships. The Plimsoll lines which designate maximum drafts allowed for vessels under various conditions are also found amidships.

Dry certificate. A document issued at the discharge port by a representative of the consignee indicating that each shipboard cargo tank has been completely discharged.

Drydock or d/d. An out-of-service period during which planned repairs and maintenance are carried out, including all underwater maintenance such as external hull painting. During the drydocking, certain mandatory Classification Society inspections are carried out and relevant certifications are issued. Modern vessels are designed to operate for five years between dry dockings. Normally, as the age of a vessel increases, the cost and frequency of drydocking increase. After the third Special Survey, drydocks will be conducted every 2.5 years.

DTA. A deferred tax asset is an item on the balance sheet that results from overpayment or advance payment of taxes.

DTL. A deferred tax liability is a tax that is assessed or is due for the current period but has not yet been paid meaning that it will eventually come due. The deferral comes from the difference in timing between when the tax is accrued and when the tax is paid.

DWT. Deadweight Tonnage is the lifting or carrying capacity of a ship when fully loaded. This measure is expressed in metric tons when the ship is in salt water and loaded to her marks. It includes cargo, bunkers, water, lubricants, stores, passengers and crew.

EBITDA. Stands for Earnings Before Interest, Taxes, Depreciation and Amortisation and is a metric used to evaluate a company's operating performance. It can be seen as a proxy for cash flow. In finance, the term is used to describe the amount of cash (currency) that is generated or consumed in a given time period.

EEDI. Energy Efficiency Design Index. The EEDI for new ships is the most important technical measure and aims at promoting the use of more energy-efficient (less polluting) equipment and engines. The EEDI requires a minimum energy efficiency level per capacity mile (e.g., tonne mile) for different ship types and size segments. Since 1 January 2013 new ship design needs to meet the reference level for their ship type.

EEOI. The Energy Efficiency Operational Index is the amount of CO_2 emitted by the ship per tonne-mile of work. It is the ratio of the CO_2 emitted to the tonne mile (amount of cargo × nautical miles sailed). The total operational emissions to satisfy the transport work demanded are usually quantified over a period of time which encompasses multiple voyages. It measures the ratio of a ship's carbon emissions per unit of transport work.

EEXI. Energy Efficiency Existing Ship Index describes, in principle, the CO_2 emissions per cargo tonne and mile. It determines the standardised CO_2 emissions related to installed engine power, transport capacity and ship speed. The EEXI is a design index, not an operational index. The EEXI is applied to all oceangoing cargo and passenger vessels above 400 gross tonnages.

EIA. The US Energy Information Administration is the statistical agency of the Department of Energy. It provides policy-independent data, forecasts and analyses to promote sound policy-making, efficient markets and public understanding regarding energy and its interaction with the economy and the environment.

Ensign. The flag carried by a ship as insignia of her nationality.

ETA. Estimated Time of Arrival.

Even keel. The existing conditions of a vessel whose fore and aft drafts are equal.

Filling density. The ratio of the weight of liquid in a tank to the weight of distilled water at 15°C (60°F) the tank will hold. It is expressed as a percentage (%).

Fixing. This is the term used for concluding a charter party negotiation. A charter party contract is often referred to as a "fixture." A vessel whose next employment has been arranged is referred to as "fixed."

Flag State. Any State that allows ships to be registered under its laws.

Flags of necessity (or convenience). Flag States that provide lesser economic, financial, tax and/or regulatory burdens to shipowners registering their ships in those countries.

Flame screen (or arrester). A device comprising a fine wire gauze that is fitted into the discharge end of a vent line. It prevents the passage of flame but will allow vapour to pass through. Flame screens are also fitted to removable ullage plugs used to cover ullage holes on cargo tank tops.

Force majeure. Clause permitting contract to be broken in the event of uncontrollable events, for example, war, strike government action, which preclude its fulfilment.

Fore, forward. Towards the stem or the bow. The section of the vessel between the stem and amidships.

Forepeak. The narrow extremity of the vessel's bow. Also, the tank located in that part of the ship.

FPSO. Stands for Floating Production, Storage and Offloading. FPSOs are designed to receive all of the hydrocarbon fluids pumped by nearby offshore platforms (oil and gas), to process it and to store it. FPSOs are typically moored offshore ship-shaped vessels, with processing equipment, or topsides, aboard the vessel's deck and hydrocarbon storage below, in the hull of the vessel.

Frames. The ribs of a ship.

Free onboard (FOB). Incoterm. The charterer is responsible for the cost of loading the cargo.

Freeboard. The distance from the water line to the top of the weather deck on the side.

Freight rate. The charge made for the transportation of freight.

FSO. A Floating Storage and Offloading vessel are commonly used in oil fields where it is not possible or efficient to lay a pipeline to the shore. The production platform will transfer the oil to the FSO where it will be stored until a tanker arrives and connects to the FSO to offload it.

Fuel oil. A name given to the heaviest grades of residual fuel used in marine oil-burning boilers.

Gangway (gangplank). A device by which persons come onboard or disembark the vessel.

Gas free. An atmospheric condition in a tank when it is free from any concentration of inflammable, noxious or toxic gases and vapours.

Gas free certificate. A certificate issued by a chemist after sampling the air in a tanker's cargo tanks after the cargo has been pumped out. It is endorsed with one of the following notations: (1) Safe for men, (2) Not safe for fire, (3) Safe for men and fire and (4) Not safe.

Gauging. A process of measuring the height of a liquid in a storage tank usually using a weighted graduated steel tape and bob.

GEI. The Bloomberg Gender Equality Index tracks the performance of public companies committed to disclosing their efforts to support gender equality through policy development, representation and transparency.

General arrangement plan. A drawing of a ship which lists all necessary statistics and operating information such as LOA, SDWT, cargo, water and fuel capacity. The deadweight scale is also contained on this important chart which is usually posted outside the ship's office or mate's cabin.

General average. A general contribution of money paid by all parties concerned in a marine adventure in direct proportion to their several interests when a voluntary or deliberate sacrifice has been made of one or more of the party's goods in time of peril with a view to saving the remainder of the property.

GHG. Green House Gas. Greenhouse gases are compound gases that trap heat or longwave radiation in the atmosphere. Their presence in the atmosphere makes the Earth's surface warmer. The principal GHGs, also known as heat-trapping gases, are carbon dioxide, methane, nitrous oxide and the fluorinated gases.

Green passport. The green passport contains details of all materials, especially which are harmful to human health, used in the construction of a vessel. The green passport will be delivered by the shipyard during the construction, and it will be later updated with all the changes made to the ship during its lifetime.

Gross standard volume (GSV). The total volume of all petroleum liquids, sediment and water, excluding free water, corrected by the appropriate volume correction factor (Ctl) for the observed temperature and API gravity, relative density, or density to a standard temperature such as 15°C (60°F) and also corrected by the applicable pressure correction factor (Cpl) and meter factor.

Gross tonnage (GT or GRT). A measurement of volume including most of the confined spaces onboard a vessel. The figure is often used as a basis for calculating port charges.

Handy size. Tankers of about 12,000–25,000 DWT.

Handymax. Tanker of about 30,000–50,000 DWT.

Harbour dues. Various local charges against all seagoing vessels entering a harbour, to cover maintenance of channel depths, buoys and lights. All harbours do not necessarily have this charge.

Hawse pipe, hawse. The hole in the bow through which the anchor chain passes.

Hawser. A cable used in warping or mooring the vessel.

Heating coils. Coils located in the bottom of cargo tanks that steam passes through to heat cargo. The heat lowers the viscosity of the cargo and permits easier pumping of the cargo at the discharge port. Vessels in clean service normally do not have or need heater coils as the viscosity of the clean products (with the exception of some lube oils) is high enough to permit easy pumping at atmospheric temperatures.

Hog (hogging). The condition of a vessel caused by the unequal distribution of cargo. When a vessel loads too heavily at the ends it causes an arching, or bending upward, of the hull at the midship area. This can also be caused by the vessel working in heavy seas with a large wave under the amidships section.

Hull. The watertight body of a ship or boat. The hull may open at the top (such as a dinghy), or it may be fully or partially covered with a deck.

HW. High water in port as determined by tides which might affect the amount of cargo a vessel can load.

IFRS. IFRS standards are International Financial Reporting Standards that consist of a set of accounting rules that determine how transactions and other accounting events are required to be reported in financial statements.

IGO. An intergovernmental organisation or international organisation is an organisation composed primarily of sovereign states (referred to as member states), or of other intergovernmental organisations.

IHM. The Inventory of Hazardous Materials is a list that provides ship-specific information on the actual hazardous materials present onboard, their location and approximate quantities.

IMO. The International Maritime Organisation's main task is to develop and maintain a comprehensive regulatory framework for shipping including safety, environmental concerns, legal matters, technical cooperation, maritime security and the efficiency of shipping. It was established by means of a Convention adopted under the auspices of the United Nations in 1948. https://www.imo.org/en

In class. A vessel currently meeting all the requirements of its classification society is "in class."

Indicated volume. The change in meter reading that occurs during a receipt or delivery.

Inert gas (IG). A gas used by marine tank vessels to displace air in cargo tanks to reduce the oxygen content to 8% or less by volume and thus reduce the possibility of fire or explosion. The inert gas used is usually nitrogen, carbon dioxide or a mixture of gases such as flue gas.

Inert gas system (IGS). A mechanical method of introducing inert gas into a vessel's tanks. An inert gas is one which has little or no ability to react with other gases or to heat. Examples of inert gases are nitrogen and CO_2. Shipboard inert gas systems utilise CO_2, either from flue gas sources or from inert gas generators.

Inerting. A procedure used to reduce the oxygen content of a vessel's cargo spaces to 8% or less by volume by introducing an "inert" gas blanket such as nitrogen or carbon dioxide or a mixture of gases such as flue gas.

Innage. The amount of space within a tank that is occupied by oil. Innages are sometimes called soundings or body gauges.

Inshore traffic zone. A designed area between the landward boundary of a traffic separation scheme and the adjacent coast intended for coastal traffic.

Inspector. A person assigned to determine the quantity and/or the quality of a commodity.

- **Company inspector.** A Company employee given the responsibility of determining the quantity and/or the quality of a volume of oil being moved or stored.
- **Independent inspector (cargo surveyor).** A person or organisation of persons acting independently, but on behalf of, one or more parties involved in the transfer, storage, inventory or analysis of a commodity for purposes of determining the quantity, and/or quality

of a commodity. They may also be assigned to the calibration of various measurement instruments and/or storage tanks ashore or on vessels.

Intake certificate. A document issued by the shipper indicating the amount of cargo loaded aboard the vessel as calculated from the shore tank gauges. Freight is paid on the basis of these figures.

Intermediate fuels. Light, residual type fuel oils with characteristics between bunker fuel and marine diesel fuel, typically used in motor ships. It is quoted in terms of Redwood per second. Redwood Second is the time, given in seconds, for 50 millilitres of a sample liquid to flow through a Redwood Viscometer.

International load-line certificate. A document issued by a classification society stating the minimum freeboard granted to a vessel and giving the position of the loading disc on the ship's side.

Intertanko. The International Association of Independent Tanker Owners is a trade association. It has served as the voice for independent tanker owners since 1970 on regional, national and international levels. The association actively works on a range of technical, legal, commercial and operational issues that have an influence on tanker owners and operators around the world.

IoT. The Internet of Things describes the network of physical objects – "things" – that are embedded with sensors, software and other technologies for the purpose of connecting and exchanging data with other devices and systems over the Internet. These devices range from ordinary household objects to sophisticated industrial tools.

IP. Institute of Petroleum (US).

ISGOTT. International Safety Guide for Tankers and Terminals.

Isherwood system. A method framing a vessel which employs closely spaced longitudinally with extra heavy floors spaced further apart. Most tankers use this type of framing system.

ISM (International Safety Management). The International Safety Management Code is a set of IMO regulations that ship operators and ships must comply with. The purpose of the ISM Code is to provide an international standard for the safe management and operation of ships and for pollution prevention.

ITF. The International Transport Workers' Federation is a democratic, affiliated federation recognised as the world's leading transport authority. The ITF has been helping seafarers since 1896 and today represents the interests of seafarers worldwide, of whom over 600,000 are members of ITF-affiliated unions. The ITF is working to improve conditions for seafarers of all nationalities and to ensure adequate regulation of the shipping industry to protect the interests and rights of the workers. The ITF helps crews regardless of their nationality or the Flag of their ship.

ITOPF. The International Tanker Owner Pollution Federation is a not-for-profit organisation established on behalf of the world's shipowners to promote an effective response to marine spills of oil, chemicals and other hazardous substances

Jacob's ladder. A rope ladder with wooden rounds is used for getting on or off a vessel, not at a berth. Also referred to as a pilot's ladder because of its extensive use by vessel pilots.

Jettison. The act of throwing goods or pumping cargo overboard to lighten a ship to improve stability in an emergency.

Keel. The backbone of the ship. It is a longitudinal beam or plate in the extreme bottom of a ship from which the ribs or floors start.

Knot. A unit of speed equal to one nautical mile (1.852 kilometre) per hour, approximately 1.151 miles per hour.

Laden loaded with cargo. The opposite is "in ballast."

Laden/ballast ratio. A comparison of the time the vessel spends employed compared with the time spent without a cargo, which is sometimes used as a management tool to assess performance.

Lay days. The period of time described in the charter party during which the owner must tender his ship for loading. The charterer is not obliged to start loading before the commencement of lay days. The charterer may cancel the charter if the ship does not tender prior to the expiration of lay days.

Laytime on a spot voyage. Laytime is the amount of time in port granted by the owners of the vessel under the terms of the charter party for the charterers to arrange for the loading and discharging of the cargo. The usual amount of total time allowed to charterers in the tanker trades is 72 running hours. Any excess time used will be paid for at an agreed rate. This is known as demurrage. In the tanker trades, any time saved (dispatch) will not be credited to charterers, however.

Length between perpendiculars (LBP). The length of the vessel measured between the forward part of the stem to the after part of the rudder post.

Length overall (LOA). The extreme length of the vessel measured from the foremost part to the aftermost part of the hull.

Letter of protest (LoP) or notice of apparent discrepancy (NAP). A letter issued by any participant in a custody transfer citing any condition with which the issue is taken. This serves as a written record that the particular action or finding was questioned at the time of occurrence.

Lighter. (1) General name for a broad, flatbottomed boat used in transporting cargo between a vessel and the shore. The distinction between a lighter and a barge is more in the manner of use than in equipment. The term "lighter" refers to a short haul, in connection with loading and unloading operations of vessels in harbours while the term "barge" is more often used when the cargo is being carried to its destination or over a long distance; (2) to load or discharge cargo to or from another vessel.

Lighterage. (1) Fee charged for conveying cargo by lighters or barges; (2) area where vessels normally lighter.

Lightering. Conveying cargo with another vessel known as a lighter from a ship to shore or voyage.

Lightweight. A measurement of mass representing the weight of the ship when completely empty. Usually expressed in long tonnes (2240 pounds).

Limber holes. Holes in the bottoms of stringers through which cargo flows through to the suction strums.

Limited liability. The law that permits a shipowner to restrict his liability to the value of this vessel after the accident plus the earnings for the voyage.

List. The leaning of the vessel to the port or starboard.

Lloyd's Register of Shipping (LRS). British classification society.

LNG. Liquefied natural gas has been made over millions of years of transformation of organic materials, such as plankton and algae. Natural gas is 95% methane, which is actually the cleanest fossil fuel. The combustion of natural gas primarily emits water vapour and small amounts of carbon dioxide (CO_2). This property means that associated CO_2 emissions are 30–50% lower than those produced by other combustible fuels.

LOA. Length overall. The length of the vessel from stem to stern. A Handymax may have an LOA of 180 metres (590 feet), a modern Panamax 225 metres (738 feet) and VLCCs over 300 metres (>984 feet).

Load displacement. The displacement of a vessel when it floats at its loading draft.

Load line. The maximum draft to which the vessel may load. The line on a vessel indicating the maximum depth to which that vessel can sink when fully loaded with cargo. Also known as its marks.

Load on top (LOT). It is defined as both a procedure and a practice. Load on top is the shipboard procedure of collecting and settling water and oil mixtures, resulting from ballasting and tank cleaning operations (usually in a special slop tank or tanks) and subsequently loading cargo on top of and pumping the mixture ashore at the discharge port.

Loaded passage. The passage during which the tanker is carrying cargo.

Log. An apparatus for measuring the speed of a vessel through the water. Also, an entry made in a logbook to record any event, for example, to enter in the logbook the name of a seaman and his offence and the penalty attached to it.

Long tonne. A unit of weight = 2,240 pounds or 1,106 kilos.

LOOP. The Louisiana Offshore Oil Port, a 19-mile-long underwater pipeline connecting Louisiana with offshore tankers. The LOOP provides

VLCCs and VPLUSs with an alternative method of delivering crude oil to ports rather than Caribbean transshipments and lightering.

LR1 and LR2. Abbreviations for Long Range oil tankers. Tankers with approx. 50,000–80,000 DWT (LR1) and approx. 80,000–120,000 DWT. (LR2).

LT (long tonne). Imperial weight measurement equal to 2240 pounds. A long tonne is 1.6% heavier than a metric tonne.

Lumpsum. The alternative to fixing on a "Worldscale" basis for spot charter parties, is to fix an agreed lumpsum freight amount for a specific voyage. Lumpsum fixtures are more common in the product trades than in the crude oil markets.

Marine surveyor. A duly qualified person who examines ships to ascertain their condition, on behalf of owners and underwriters. Also called a "ship surveyor" or simply "surveyor."

Maritime law. That system of jurisprudence that prevails in courts having jurisdiction of marine causes. Also called marine or admiralty law. It is a branch of both international and commercial law.

MARPOL regulations. A series of internationally ratified IMO regulations pertaining to the marine environment and the prevention of pollution.

Mbpd. Million barrels per day.

Mean draft. The average of the drafts measured at the bow and the stern.

Metric tonne. A unit of weight equal to 1,000 kilograms or 2,204.6 pounds.

Midship draft. The draft read at the midship markings. This draft can, and often does, differ from the mean draft due to hogging or sagging.

Mooring line. Any hawser by which a vessel is secured to a dock or mooring. It may be made of natural materials (manila), synthetics, (polypropylene) or wire. Under certain circumstances, the anchor chain is detached from the anchor and a section of that is used to secure the vessel.

MOPU. A Mobile Offshore Production Unit is any type of portable structure that can be reused when procuring oil and gas from the seabed. These are typically used when the depth of drilling is over 500 metres (1,640 feet). If the water is any shallower, then fixed platforms are constructed

Moulded breadth. The breadth of the hull at the widest part, measured between the outer surfaces of the frames.

Moulded depth. The depth measured between the top of the keel, or lower surface of the frame at the centreline, and the top of the upper deck beam at the gunwale.

MR (medium range). Another term for a tanker of around 25,000–50,000 DWT (see Handymax).

MT, or metric tonne. A unit of weight, one tonne being 1,000 kilos. Note that one cubic metre of fresh water (SG 1.00) at 15°C (59°F) will weigh exactly one metric tonne.

N/B. New building.

Net capacity. The number of tons of cargo which a vessel can carry when loaded in salt water to her summer freeboard marks. Also called cargo carrying capacity, cargo deadweight and useful deadweight.

Net registered tonnage. The internal capacity of a vessel measured in units of 100 cubic feet less the space occupied by boilers, engines, shaft alleys, chain lockers, officer's and crew quarters and other spaces not available for carrying passengers or freight. Net registered tonnage is usually referred to as registered tonnage or net tonnage.

Net standard volume (NSV). The total volume of all petroleum liquids, excluding sediment and water and free water, corrected by the appropriate volume correction factor (Ctl) for the observed temperature and API Gravity, relative density or density to a standard temperature such as 15°C (60°F) and also corrected by the applicable pressure correction factor (Cpl) and meter factor.

Net tonnage. The volumetric cargo capacity of a ship is expressed on the basis of 100 cubic feet per the tonne. On passenger vessels, it also includes space used by passengers.

Norske Veritas. Norwegian classification society.

Notice of readiness (NOR). Notice served by the Master to inform the terminal/charterer the vessel is ready in all respects to load or discharge cargo.

NOx. In atmospheric chemistry, NOx is a generic term for the nitrogen oxides that are most relevant for air pollution, namely nitric oxide (NO) and nitrogen dioxide (NO_2). These gases contribute to the formation of smog and acid rain, as well as affecting tropospheric ozone.

OBQ (onboard quantity). The material remaining in vessel tanks, void spaces, and/or pipelines prior to loading. Onboard quantity includes water, oil, slops, oil residue, oil/water emulsions, sludge and sediment.

OCIMF. The Oil Companies' International Marine Forum is an organisation of oil companies that own or operate ships.

OECD. The Organisation for Economic Cooperation and Development is an international organisation that works to build better policies for better lives. The goal is to shape policies that foster prosperity, equality, opportunity, and wellbeing for all.

Off-hire day. Each day, or part thereof, during which a tanker is not earning revenue from the charterer.

Off specification product/cargo. Refined products or other cargo that does not meet normal quality requirements and therefore require special handling and restraints to assure separation from specification products/cargo.

Onboard quantity (OBQ). The material remaining in vessel tanks, void spaces, and/or pipelines prior to loading. Onboard quantity includes water, oil, slops, oil residue, oil/water emulsions, sludge and sediment.

OPA90, or US Oil Pollution Act of 1990. Legislation passed in the US which details, amongst other things, certain regulations regarding the age and hull type of tankers and vessel operators' liabilities and responsibilities with regard to the US marine environment.

OPEC. Organisation of Petroleum Exporting Countries. OPEC is an organisation of 13 oil-producing countries whose members include Algeria, Gabon, Indonesia, Iran, Iraq, Kuwait, Libya, Nigeria, Qatar, Saudi Arabia, United Arab Emirates, and Venezuela. The mission of the organisation is to "coordinate and unify the petroleum policies of its member countries and ensure the stabilisation of oil markets, in order to secure an efficient, economic and regular supply of petroleum to consumers, a steady income to producers and a fair return on capital for those investing in the petroleum industry.

OPEC+. The Organisation of the Petroleum Exporting Countries Plus is a loosely affiliated entity consisting of the 13 OPEC members and 10 of the world's major non-OPEC oil exporting nations.

Operating costs. The costs incurred by the shipowner during a voyage or time charter and by the charterer in a bareboat charter. Comprising crew wages and associated costs; insurance (hull and machinery, and protection and indemnity premiums); cost of lubricants and spare parts; repair and maintenance (routine maintenance, dry dockings and classification fees).

Outage (ullage). The depth of the space in a tank not occupied by oil. Same as ullage. It is measured from the flange of the ullage hole to the surface of the oil. Also, the space left in a petroleum product container to allow for expansion as a result of temperature changes during shipment and use.

Outturn certificate. A document issued by the receivers of cargo indicating the amount of cargo discharged.

Panamax. The maximum size ship that can fit through the Panama Canal in terms of width, length and draft about 80,000 DWT.

Passage. A journey from one port or place to another, as distinguished from the term "voyage" which refers to a ballast and loaded passage. Also sometimes called a trip.

PCNT (Panama Canal Net Tonnage). A measurement based on volumes in accordance with the Panama Canal Authority rules, which is used to calculate the charges for transiting the Panama Canal.

Peak tank. Tanks in the forward and after ends of the vessel. The principal use of peak tanks is in trimming the ship.

Per calendar day (month/year) costs. Vessel's costs expressed in US Dollars (US$) per day (month/year) for a calendar period during which the vessel was in service. The number of calendar days (months/years) is divided into the total costs incurred during the period.

Per operating day (month/year) costs. Vessel's costs expressed in US Dollars (US$) per day (month/(year) during which the vessel actually operated. It includes the costs incurred while the vessel was idle for repairs or other nonoperating reasons. The number of operating days (excluding nonoperating delays) is divided into the total costs.

Pilot house. The enclosed space on the navigating bridge from which a ship is controlled when underway.

Plating. The steel plates which form the shell or skin of the vessel.

Plimsoll line. A reference mark located on a ship's hull that indicates the maximum depth to which the vessel may be safely immersed when loaded with cargo. This depth varies with a ship's dimensions, type of cargo, time of year and the water densities encountered in port and at sea.

Pool. A pool is a group of similar size and quality vessels with different shipowners that are placed under one administrator or manager. Pools allow for scheduling and other operating efficiencies such as multilegged charters and contracts of affreightment.

Pool points. A system of pool points creates a model for a vessel with a performance equating to the average of those being pooled. This ship is awarded 100 pool points. All other ships in the pool are then given more or less pool points adjusted for the characteristics of each vessel. Pool points, by their nature, can only be used to address the differences between the vessels as described and not the vessel as performed.

Port. The left side of a vessel when an observer is facing forward looking towards the bow. Also, a door on a ship.

Port charges. General term which includes charges and dues of every nature assessed against the vessel or its cargo in a port. It usually includes harbour dues, tub boat charges, pilotage fees, custom house fees and consular fees.

Port of registry. The port at which a vessel is registered and to which she is considered to belong. The port of registry is shown on the stern below the name of the vessel.

Port state. A state that has ports to which ships call. The port state makes regulations the calling ships must adhere to. The port state control is the controlling authority of the port state on shipping such as the coast guard or naval authorities.

Port time (two types). (1) Sea buoy to sea buoy. The time elapsed between the vessel's passing the port's sea buoy upon entrance to repassing it upon exit. It includes time for steaming in and out of berth, delays, hose connections, anchorage time, clearing and loading or unloading time; and (2) port to port. Includes only time for delays, hose connections, anchorage time, clearing and loading or unloading time. Use must be consistent with voyage mileage basis.

Portable measurement unit (PMU). A device designed to measure the ship's cargo when its tanks are closed to the atmosphere. It is used in conjunction with a vapour control valve.

Portable Sampling Unit (PSU). A device designed to sample the ship's cargo when its tanks are closed to the atmosphere. It is used in conjunction with a vapour control valve.

Position report (position sheets). A summary of worldwide movements for vessel prepared by the Fleet Coordinators.

Posted price (contract price). The price for marine fuel oils which appears on a price list published by marine fuel oil (bunker) brokers.

Pour point. The lowest temperature at which oil will remain liquid.

Practice. Load on top is the act of commingling onboard quantity with cargo being loaded.

Premium. Surcharge over general market rate level to compensate the vessel's owner for an unusually difficult trade, for example, Lake/Aruba, or to correct for an imbalance in supply/demand conditions in a given area.

Pressure. The amount of force exerted on a unit of area by a fluid.

- **Absolute pressure.** The pressure referenced to a perfect vacuum as zero pounds per square inch absolute (psia).
- **Atmospheric pressure.** The pressure exerted by the atmosphere. Although this pressure varies with altitude, barometric pressure and humidity, the atmospheric pressure can be defined in custody transfer contracts, or by state and federal authorities. Atmospheric pressure is most often stated as 14.696 pounds per square inch absolute.
- **Back pressure.** The operating pressure level measured upstream from a control valve.
- **Gauge pressure.** That pressure measured relative to atmospheric pressure as zero, usually designated psig.
- **High vapour pressure.** A fluid which, at the measurement or proving temperature, has a vapour pressure that is equal to or higher than atmospheric pressure.
- **Low vapour pressure.** A fluid which, at the measurement or proving temperature, has a vapour pressure that is less than atmospheric pressure.
- **Reid vapour pressure.** The vapour pressure of a fluid at 37.7°C (100°F) as determined by test method ASTM D 32358.
- **Static pressure.** The pressure in a fluid that is exerted normally to the surface. In a moving fluid, the static pressure is measured at right angles to the direction of flow.

Pressure/vacuum valve (P/V valve). An automatic dual-purpose valve, commonly fitted in the vent lines of tankers. When in the closed position, the function of this valve is to relieve either pressure or vacuum in a tank. When in the open position it allows the passage of air or vapour into and out of the tank.

Protection and indemnity (P&I) insurance. Mutual protection provided by an association of shipowners against liabilities not covered by insurance.

Protest, Notice of. A letter issued by any participant in a voyage citing any condition with which the issue is taken. This serves as a written record that the particular action or finding was questioned at the time of occurrence. For example, a declaration made by the Master before a notary public or consular official when through stress of weather, there has been or the Master fears that there might have been, damage to the vessel or cargo. Copies are frequently demanded by insurance underwriters in the event of a claim.

Pump room. An enclosed area on a tank vessel which houses main and stripping cargo pumps, ballast pumps, educators and the associated piping and valves necessary for their operation.

Quarter. A side of a ship aft, between the main midship frames and stern. Also, a side of the ship forward, between the main frames and stem.

Rate. The cost, or revenue, for a particular voyage based on a standard of reference, for example, Worldscale, INTASCALE and ATRS.

Reducer. A short section of pipe, having one end of smaller diameter than the other and having a flange on each end, for connecting a smaller hose or pipe to a pipe of constant diameter.

Registry A duty imposed on shipowners in order to secure to their vessels the privileges of ships of the nation to which they belong.

Relative density. See Density.

Remaining onboard (ROB). The material remaining in vessel tanks, void spaces and/or pipelines after discharge. Remaining onboard quantities includes water, oil, slops, oil residue, oil/water emulsions, sludge and sediment.

Repositioning. The movement of a vessel in ballast to shift it from one trading pattern to another.

Restricted measurement system. A measurement system designed to measure the ship's cargo when its tanks are closed to the atmosphere. During measurements, a minimum amount of cargo vapours might escape to the atmosphere.

ROB (Remaining onboard). The material remaining in vessel tanks, void spaces and/or pipelines after discharge. Remaining onboard quantities includes water, oil, slops, oil residue, oil/water emulsions, sludge and sediment.

Rogue wave. An ocean wave much larger than the current wave sequence. This wave may also be outside the current wave direction and may be 30 metres (100 feet) or more in height.

Route. See Deep water route, traffic route and two-way route. Means whichever type is appropriate in the context unless otherwise specified.

Routing. A complex of measures concerning routes aimed at reducing the risk of casualties; it includes traffic separation schemes, two-way routes, tracks, areas to be avoided, inshore traffic zones and deepwater routes.

Rudder. The flat or shaped frame hung to the sternpost of a ship, which is used to steer the ship.

Rules of the road. The rules and regulations accepted by international agreement and enforced by law in marine countries which govern the movements of ships when approaching each other under such circumstances that a collision may ensue.

Safe for men. A term signifying that the vapour content of a space so certified is less than 0.1 on a gas indicator.

Safe for men and fire. A term signifying that the vapour content of a space so certified is 0.1 or less on a gas indicator and that the space contains no oil or sediment which could produce vapours.

Sag (sagging). The condition of a vessel caused by the unequal distribution of cargo. When a vessel loads too heavily in the centre it causes a bending downward of the hull at the midships area. This can also be caused by the vessel working in heavy seas with large waves under each end and no support under the centre of the ship. Sag is the opposite of Hog.

Salvage. The property which has been recovered from a wrecked vessel, or the recovery of the vessel herself.

Sampling. The process of obtaining a sample of the material in the tank, container or pipeline to use for testing or other purposes. This can be achieved by automatic or manual means. The following are the most common types of samples taken:

- **All levels sample.** A sample obtained by lowering a weighted, stoppered bottle or beaker or bottle to a point 1 foot (0.3 metre) above the free water level and then, with a sharp jerk of the line opening the sampler and raising it at a rate that it is about 75% full (a maximum of 85% full) as it emerges from the liquid.
- **Automatic sample.** A sample taken by automatic means. The two basic types of automatic samples are:
- **Bottom sample.** A spot sample taken from the material at the bottom of the tank.
- **Flow proportional sample.** A sample taken by an automatic sampler from a pipeline at a rate that is proportional to the liquid flow rate.
- **Lower sample.** A spot sample obtained at the midpoint of the lower third of the tank contents.
- **Middle sample.** A spot sample obtained at the midpoint of the middle of the tank contents.
- **Running sample.** A sample obtained by submerging an unstoppered beaker or bottle from the surface of the liquid to a point as near as possible to the shore tank draw-off point or about one foot above

the level of the free water in a ship tank, and then raising it without letting it rest, at a rate so that it will be about 75% full as it emerges from the liquid.

- **Spot sample.** A sample is taken at a specific "spot" within a tank using a stoppered bottle or beaker and lowering it to the level of the desired sample then opening it and allowing it to remain at that level until full. A thief or a zone sampler may also be used to obtain spot samples.
- **Tap sample.** A sample is taken from a valve or connection on a tank or pipeline.
- **Time proportional sample.** A sample in taken from a pipeline at regular intervals during a batch transfer period.
- **Upper sample.** A spot sample was obtained at the midpoint of the upper of the tank contents.
- **Upper, middle, lower samples.** Spot samples taken from the upper third, the middle and lower thirds of the liquid in the tank. The samples so taken may then be composited or analysed separately.

SBT. Segregated ballast tanks are dedicated tanks constructed for the sole purpose of carrying ballast water on oil tanker ships. They are completely separated from the cargo, and fuel tanks and only ballast pumps are used in the SBT.

Scrubbers. Shortened term for Exhaust Gas Cleaning Systems (EGCS), or SOx (sulphur dioxide) scrubbers. These are used to remove harmful elements (sulphur oxides) from exhaust gases from vessels by using wash water from the sea to neutralise the exhaust product. There are two key categories open loop scrubbers which discharge wash water used into the ocean and closed loop which retain the waste product until it can be delivered to an appropriate location.

Scupper. Any opening or tube leading through the ship's side to carry water away from the deck.

SDG. The Sustainable Development Goals, also known as the Global Goals, were adopted by all United Nations Member States in 2015 as a universal call to action to end poverty, protect the planet and ensure that all people enjoy peace and prosperity by 2030.

Sea trials. A series of trials conducted by the builders during which the owner's representatives onboard act in a consulting and checking capacity to determine if the vessel has met the specifications.

Seaworthiness. The sufficiency of a vessel in materials construction, equipment, crew and outfit for the trade in which it is employed. Any sort of disrepair to the vessel by which the cargo may suffer; overloading; untrained officers; may constitute a vessel unseaworthy.

Seaworthiness certificate. A certificate issued by a classification society surveyor to allow a vessel to proceed after she has met with a mishap that may have affected its seaworthiness. It is frequently issued to enable a vessel to proceed, after temporary repairs have been effected, to another port where permanent repairs are then carried out.

SEEMP. The Ship Energy Efficiency Management Plan is an operational measure that establishes a mechanism to improve the energy efficiency of a ship in a cost-effective manner. The SEEMP also provides an approach for shipping companies to manage ship and fleet efficiency performance over time using, for example, the Energy Efficiency Operational Indicator (EEOI) as a monitoring tool.

Separation zone or line. A zone or line separating traffic proceeding in one direction from traffic proceeding in another direction. A separation zone may also be used to separate a traffic lane from the adjacent inshore traffic zone.

Shale oil. Crude oil that is extracted from oil shale (fine-grained sedimentary rock containing kerogen) by using techniques other than the conventional (oil well) method, for example, heating and distillation.

Ship chandler. Particular merchants handling ship's stores and supplies and sundries. Sometimes handle spare parts as accommodation to ship operators.

Shipbreaker. A company that demolishes or cuts up vessels which are obsolete or unfit for sea. The steel is used for scrap.

Shipper. The person for whom the Master of a ship agrees to carry cargo. Also called the consignor.

Ship's agent. A person or firm who transacts all business in a port on behalf of shipowners or charterers. Also called shipping agent, agent.

Short tonne. A unit of measurement equal to 2,000 pounds.

Sister ships. Ships built on the same design.

Skin. The plating of a ship.

Slops. A mixture of petroleum and water normally arising from tank washings.

Sludge. A mixture of petroleum and water, usually semisolid, frequently containing sand and scale.

SOLAS. Safety of Life at Sea Convention.

Sounding. See Gauging.

SOx. The two main pollutants from the ship's emission are nitrogen oxides (NOx) and sulphur oxides (SOx). These gases have adverse effects on the ozone layer in the troposphere area of the earth's atmosphere which results in the greenhouse effect and global warming.

Spar. A single-point mooring and reservoir is a type of floating oil platform typically used in very deep waters and is named for logs used as buoys in shipping that are moored in place vertically. Spar production platforms have been developed as an alternative to conventional platforms.

Special survey. The survey required by the classification society that usually takes place every five years and usually in a drydock. During the special survey, all vital pieces of equipment and compartments and steel structures are opened up and inspected by the classification surveyor.

Spill. Oil getting into the sea in any amount for any reason.

Spot charter. See voyage charter.

Spot market. The market for the immediate charter of a vessel.

Spot price. The current market price for an asset or commodity.

Starboard. The right side of a vessel when an observer is facing forward looking towards the bow.

Stem. The upright post or bar of the bow.

Stern. The aftermost part of a vessel. The stern will house the steering gear room and various stowage areas. It is that section of a vessel over the rudder and propeller.

Stores. A general term for provisions, materials and supplies used aboard ship for the maintenance of the crew, and for the navigation, propulsion and upkeep of the vessel and its equipment.

Submarine loading terminal. A terminal where loading is carried out by means of an offshore hose run along the sea bottom.

Suezmax carrier. The maximum size vessel that can sail loaded through the Suez Canal. This is considered to be between 120,000 and 199,999 DWT and mostly about 150,000 DWT, depending on a ship's dimensions and draft. These tankers can transport up to one million barrels of crude oil.

Super slow steaming. Reducing operating speeds in order to save fuel. Operating laden speeds are reduced from 15 knots to about 13 knots and operating ballast speeds from 15 knots to about 10–8 knots.

Superstructure. Any structure built above the uppermost complete deck such as a pilothouse, bridge and accommodation.

Sustainability-linked loan. Sustainability-linked Loans or ESG Linked Loans are general corporate-purpose loans used to incentivise borrowers' commitment to sustainability and to support environmentally and socially sustainable economic activity and growth. Under this lending model, borrowers pay higher interest rates when they fail to meet certain environmental, social and governance-linked goals. Similarly, they pay less when they exceed ESG targets.

Tackle. Any combination of ropes and blocks that multiply power. The equipment on a vessel used to perform working tasks on the vessel.

Tank washing. The cleaning of a vessel's tanks. It is divided into two types of activities:

Water washing. The use of a high-pressure water stream to dislodge clingage and sediment from the bulkheads, bottom and internal tank structures of a vessel.

TCE. Time Charter Equivalent rate is a standard shipping industry performance measure used primarily to compare period-to-period changes in a shipping company's performance despite changes in the mix of charter types (i.e., spot charters, time charters and bareboat charters) under which the vessels may be employed between the periods. A standard method to compute TCE is to divide voyage revenues (net of expenses) by available days for the relevant time period. Expenses primarily consist of port, canal and fuel costs.

Technical management. The management of the operation of a vessel, including physically maintaining and repairing the vessel, maintaining necessary certifications and supplying necessary stores, spares and lubricating oils. Responsibilities also generally include selecting, engaging and training crew and could also include arranging necessary insurance coverage.

Tender (TO). A Master tenders his vessel when he advises the charterer or supplier that he is ready to load.

Thieving. Determining the amount of water at the bottom of a tank of oil.

Time charter. A charter for a fixed period of time, usually between one and ten years, under which the owner hires out the vessel to the charterer fully manned, provisioned and insured. The charterer is usually responsible for bunkers, port charges, canal tolls and any extra cost related to the cargo. The charter rate (hire) is quoted in terms of the total cost per day. Subject to any restrictions in the charter, the customer decides the type and quantity of cargo to be carried and the ports of loading and unloading.

TLP. A tension leg platform or extended tension leg platform (ETLP) is a vertically moored floating structure normally used for the offshore production of oil or gas and is particularly suited for water depths greater than 300 metres (about 1,000 Annual report 2020 132 feet) and less than 1,500 metres (about 4,900 feet). The use of tension leg platforms has also been proposed for wind turbines.

Tonnage. See deadweight, GRT and NRT.

Tonnage tax. Taxes, fees and harbour dues paid on the vessel based on a tonnage calculation.

Tonnage tax regime. An alternative way of calculating the taxable income of operating qualifying ships. Taxable profits are calculated by reference to the net tonnage of the qualifying vessels a company operates, independent of the actual earnings (profit or loss).

Tonne. Typical unit of weight measurement used on tankers. See long tonne, metric tonne and short tonne.

Tonne. Metric tonne.

Tonne-mile. A unit for freight transportation equivalent to a tonne of freight moved one mile.

Tonne mile. Quantity transported multiplied by average voyage distance. Also, Ton-mile.

Tonne-mile demand. A calculation that multiplies the average distance of each route a tanker travels by the volume of cargo moved. The greater the increase in long-haul movement compared with shorter haul movements, the higher the increase in tonne-mile demand.

Tons per inch immersion (TPI). The number of tons required to change a vessel's draft one inch in the water. TPI varies with the draft and its values can be found on a vessel's deadweight scale. The metric equivalent is known as TPC (Tons per Centimetre).

Topping-off. (1) The operation of completing the loading of a tank to a required ullage; (2) filling up cargo tanks which were only partially filled at the loading port because of port or canal draft restrictions. The filling up occurs outside the loading port via lightering activities or at another loading port.

Total calculated volume (TCV). The total volume of all petroleum liquids and sediment and water, corrected by the appropriate volume correction factor (Ctl) for the observed temperature and API gravity, relative density or density to a standard temperature such as 15°C (60°F) and also corrected by the applicable pressure factor (Cpl) and meter factor, and all free water measured at observed temperature and pressure (gross standard volume plus free water).

Total observed volume (TOV). The total measured volume of all petroleum liquids, sediment and water and free water at observed temperature and pressure. Note, where the term 15°C (60°F) is used, it refers to two different reference standards and does indicate that the two temperatures are equal.

Trim. The condition of a vessel with reference to its longitudinal position in the water. It is the difference between the forward and after drafts expressed in feet/inches or metres/centimetres. Trim forward is called "by the head" and trim aft is called "drag."

Trim by the head (by the stern). A vessel is said to trim by the head (or stern) when its draft forward (or aft) is greater than aft (or forward).

Ullage. See Outage

Ullages. Measurements taken with a steel tape from the lip of the ullage hole to the to the surface of the liquid; usually read to the nearest 1/ 8 inch.

Underwriter. In marine insurance one who subscribes his name to the policy indicating his acceptance of the liability mentioned therein consideration for which he receives a premium.

Unseaworthiness. The state or condition of a vessel when it is not in a proper state of maintenance, or if the loading equipment or crew, or in any other respect is not ready to encounter the ordinary perils of sea.

US Calls. Letters begin with "K." Liberian registered vessel callsigns begin with the letters "A" or "E" or a numeral. Call letter must be used in cables after a vessel's name.

US Gulf. The Gulf of Mexico.

Used laytime. The amount of laytime that was taken by the vessel for loading and discharging on a voyage.

Vapour control valve (VCV). This valve is used in conjunction with closed and restricted measurement equipment to allow measurements in the ship's tanks that are closed to the atmosphere. Once a portable measurement unit (PMU) is attached to the VCV, the valve is opened and the PMU's probe is dropped into the tank to perform the required measurements.

Vapour recovery system (VRS). Procedures and equipment for the collection of hydrocarbon vapours from the vessel's tanks and the transfer to shoreside recovery equipment.

Vessel expenses. Includes crew costs, vessel stores and supplies, lubricating oils, maintenance and repairs, insurance and communication costs associated with the operation of vessels.

Vessel experience factor (VEF). A factor based on the compilation of the history of the total calculated volume (TCV) vessel measurements, adjusted for onboard quantity (OBQ) or remaining onboard (ROB), compared with the TCV shore measurements. This factor if developed according to the latest industry standards may be used to obtain a better ship/shore comparison of volumes.

Vessel sizes and uses. Tankers and barges come in all sizes from the small harbour/lake variety to the biggest things ever built by man that move. The size of any particular tanker depends on many factors. Use, cargo type, amount and demand, passage length and port restrictions at both the load port and the discharge port are among the most important of these.

- **Aframax.** A tanker of 80,000–120,000 DWT.
- **Panamax.** A tanker whose dimensions enable her to transit the Panama Canal where lock width is the determining factor. Ships are 55,000–79,999 DWT.
- **Suezmax.** A tanker whose dimensions enable her to transit fully loaded through the Suez Canal. Ships are 120,000–199,999 DWT.
- **VLCC (very large crude carrier).** 200,000–320,000 DWT
- **VPLUS (ultra large crude carrier).** 320,000 DWT and above. Because of their huge sizes, these vessels have been exclusively only used for the carriage of crude oils. Only the smallest of this category has carried any type of refined products. Several of these VPLUS classed vessels were over 500,000 and the biggest of these ships had a deadweight of 564,939 tons.

Vetting. The general process of approving a vessel for use. (From old English "To Vet" to look at or review again.) Note: actual procedure varies from company to company.

Vetting. Ship Vetting is a risk assessment process carried out by charterers and terminal operators in order to avoid making use of deficient ships or barges when goods are being transported by sea or by inland waterways.

VLCC. The abbreviation for Very Large Crude Carrier. Tankers with a capacity between 200,000 and 320,000 DWT. These tankers can transport up to two million barrels of crude oil. See also Vessel sizes.

VLCC equivalent. The capacity of one VLCC or two Suezmax vessels.

Volume. The amount of space occupied by a fluid at certain conditions of temperature and pressure. Various types of volumes used in marine custody transfer are defined as the Gross Observed Volume (GOV), which is the total volume of all petroleum liquids, sediment and water, excluding free water, at observed temperature and pressure.

Voyage charter. A contract in which a charterer pays a shipowner for the use of a ship's cargo space for one, or sometimes more than one, voyage. The shipowner is the operator, responsible for paying both operating costs and voyage costs. Also referred to as a spot Charter.

Voyage costs. Voyage expenses, which include fuel, port charges, canal tolls, cargo handling operations and brokerage commissions paid by the shipowner under Voyage Charters. These expenses are subtracted from shipping revenues to calculate Time Charter Equivalent revenues for Voyage Charters.

VPlus. A crude oil tanker (ULCC or Ultra Large Crude Carrier) of more than 350,000 DWT which makes it one of the biggest oil tankers in the world. These tankers can transport up to three million barrels or more of crude oil and are used on the same long-haul routes as VLCCs. To differentiate them from smaller ULCCs, these ships are sometimes given the VPlus size designation.

VPLUS/ULCC. Ultra Large Crude Carriers are the largest shipping vessels in the world with a size ranging between 320,000 and 500,000 DWT. Due to their mammoth size, they need custom-built terminals. As a result, they serve a limited number of ports with adequate facilities to accommodate them. They are primarily used for very long-distance crude oil transportation from the Persian Gulf to Europe, Asia and North America. ULCC are the largest shipping vessels being built in the world with standard dimensions of 415 metres (1,361 feet) in length, 63 metres (206 feet) in width and 35 metres (114 feet) in draft.

VTBL. Vessel to be lightered.

Wall wash test. The procedure of introducing an appropriate liquid into a vessel's tank to test for hydrocarbon, colour and other contaminants. This test is done by physically pouring the liquid down the vessel's tank bulkheads and trapping a portion on filter paper. This test is also done on the vessel's steam coils and sumps.

Water/cut measurement. The procedure of locating the oil/water interface for the purpose of determining the volume of free water in a shore tank or vessel compartment. It is also used to refer to the line of demarcation of the oil/water interface.

Watertight door. A door so constructed that, when closed, it will prevent water under pressure from passing through.

Wedge formula. A mathematical means to approximate small quantities of liquid and solid cargo and free water onboard prior to loading and after discharge based on cargo compartment dimensions and vessel trim. The wedge formula is to be used only when the liquid does not touch all bulkheads of the vessel's tanks.

Wedge table. A precalculated vessel table based on the wedge formula and displayed much like the vessel's usual innage/ullage tables. These tables, however, are for small quantities (onboard quantities, remaining onboard) when the cargo or free water does not touch all bulkheads of the vessel tank.

Wing tanks. Vessel tanks located at the port or starboard of the centreline and designated port or starboard wings or wing tanks.

Wipe test. The procedure of physically wiping random interior areas and steam coils of vessel's tanks with absorbent white rags. This procedure is used to test the tank's coating for colour contamination.

Worldscale. The New Worldwide Tanker Nominal Freight Scale is a catalogue of theoretical freight rates expressed as US dollars per tonne for most of the conceivable spot voyages in the tanker trade. The final rate agreed will be determined as a percentage of the "Worldscale" rate, based upon a guaranteed minimum quantity of cargo. This allows for charter parties to cover a wide range of voyage options without the need to calculate and negotiate each one separately.

WTI oil price (US Oil). West Texas Intermediate, one of three main benchmarks for oil pricing.

Other Routledge publications by the same author

Olsen, Alexander. 2022. *Core Principles of Maritime Navigation*. Routledge, London.
Olsen, Alexander. 2022. *Introduction to Container Ship Operations and Onboard Safety*. Routledge, London.
Olsen, Alexander. 2023. *Firefighting and Fire Safety Systems on Ships*. Routledge, London.
Olsen, Alexander. 2023. *Introduction to Ship Engine Room Systems*. Routledge, London.
Olsen, Alexander. 2023. *Maritime Accident and Incident Investigation*. Routledge, London.
Olsen, Alexander. 2023. *Maritime Cargo Operations*. Routledge, London.
Olsen, Alexander. 2023. *Merchant Ship Types*. Routledge, London.

Index

Note: Page locators in **bold** and *italics* represents tables and figures, respectively.